THE LAST MEN OUT

# THE LAST MEN OUT

### LIFE ON THE EDGE AT RESCUE 2 FIREHOUSE

## TOM DOWNEY

A HOLT PAPERBACK

**HENRY HOLT AND COMPANY, NEW YORK**

Holt Paperbacks
Henry Holt and Company, LLC
*Publishers since 1866*
175 Fifth Avenue
New York, New York 10010
www.henryholt.com

Library of Congress Cataloging-in-Publication Data
Downey, Tom.
    The last men out : life on the edge at Rescue 2 firehouse / Tom Downey.—1st ed.
    p. cm.
    ISBN-13: 978-0-8050-7844-2
    ISBN-10: 0-8050-7844-4
    1. Rescue 2 (Fire department : New York, N.Y.)—History.   2. Fire extinction—
New York (State)—New York—History.   I. Title.
TH9505.N5071D69 2004
363.37'09747'1—dc22

Henry Holt books are available for special promotions and premiums.
For details contact: Director, Special Markets.

Originally published in hardcover in 2004 by Henry Holt and Company

First Holt Paperbacks Edition 2005

DESIGNED BY CATHRYN S. AISON

Printed in the United States of America

D   24

*To the firemen I have known, living and dead,*
*especially my father, Tom, my Uncle Ray, my Uncle Gene,*
*and the men of*
*Rescues 1 and 2*

# CONTENTS

# THE LAST MEN OUT

# PROLOGUE

THE LATE PATTY BROWN, an FDNY captain who had served as a fireman in Rescue 2, told me this story back in the summer of 2001:

*You could tell by the guy's voice that something was going on. We were ordered to respond to Seventh Avenue and Forty-eighth Street. They're starting to say on the radio that there's a bad fire—people are trapped. So we get there, and two people are hanging out different windows. We started charging up the stairs. It was a twelve-story building. I was so pumped I don't even remember that being a hindrance. I've been on the job a long time, and I've been around real tragedy—situations where our guys were killed. Even if you don't do anything wrong, you can get killed on this job. I'm always trying to be ready for whatever can happen.*

*I ran up to the roof because I wanted to see where these guys were trapped. There was nobody up there, but they were constructing a building right across the street and these construction workers were screaming and pointing below me. I had to climb up on the parapet to see where this guy was. There he was, right under me, hanging from the window ledge, ready to jump. The fire was roaring right next to him, so I ran back to the stairwell and I told the others to get up there.*

"Hold on," I screamed to the victim. He looked up at me as if he was braced to jump. "We're coming to get you," I said. He wasn't responsive. He just looked.

Pat Barr was carrying the lifesaving rope. It's always been a tradition everywhere I know that if you're carrying the rope, you're the guy they're going to lower. For safety, we take a separate rope, and we tie it to the guy who's lowering the rescuer. He's attached to an object on the roof so he doesn't go over.

Now we realize we don't have anywhere substantial to tie this rope the way we're supposed to. This guy was going to jump any minute, any minute. Even thinking of it, I get all upset. As a lieutenant, I could have said that we aren't doing this. And the guy would have died. I could have said that it's too dangerous, and nobody would have said a word. If I had finagled around and said, "Break the wall so we can tie off and have a safety line," he would have been gone. The victim was four or five feet below, looking up at me. It was either let's do it or not.

I just said, "I know we can do it. Here's what we can do: Kevin, brace yourself against the wall. We're going to hold you down." So we tied Pat Barr to Kevin, and Pat went over on the rope. Now Pat's hanging thirteen flights above the street, attached only to Kevin, and we're slowly lowering him. I was scared to death. In 1980, a guy from my firehouse was killed from the same thing. He had rescued another firefighter, and the rope snapped and they fell to their death. That guy, Larry Fitzpatrick, was a good friend of mine. I had been a fireman three or four years when he got killed. Now here I was directing the same thing.

This victim was on the precipice of life. It was seconds. Pat Barr grabbed him, and that dead weight stretched the rope so intensely that Kevin, who was lowering Pat from the roof, started pinioning right up. At this point, Paddy O'Keefe jumped on top of Kevin to hold him down, anchoring him to the roof. We lowered Barr and the victim to the floor below, *and they're just hanging there outside the window. I'm screaming on the radio for somebody. The fire is roaring ten feet from them. None of our guys on that floor below the fire knew they were there. If the fire had come out, it would have burned the rope.*

*Pat swung around and broke the windowpane with his hand, while still holding on to the victim. A fireman saw them, and the guys all ran over and pulled them in. The construction workers started cheering. Everybody's going, "All right!" But I remember saying to Paddy, "There's another guy trapped on the other side."*

*As we came to where the other guy was trapped, another fireman from Ladder 24, this guy Ray McCormack, came up to the roof. All I said to him was, "You got a belt to lower him?" He said yeah. I told him, "Okay, you're going to lower Kevin." And we did exactly the same thing as the first one. We saved both these guys. I was elated. But there was no rah-rah to it. I was humbled.*

*I had nightmares for weeks about what could have happened. There are so many unknowns in this job. Sometimes I wonder. It worked out great; it worked out fine. But even now I think: Man, oh man, that was such a close call. It was the right thing to do, but it was right on the edge.*

When Patty told me that story, his words transported me and made me feel a little of what he must have felt when he sent his men over the edge of that roof. Stories like this are staples at firehouse kitchen tables. That particular story became a legend. If you ask anybody in the FDNY about Patty Brown, they'll tell you that story. New York firefighters live in a culture kept alive by a rich tradition of storytelling. Almost none of these stories are written down. If you look in an old newspaper to find out the details of a blaze, the news story usually contains only scant detail about what the firemen actually did at the fire. But ask a fireman who knows about the same blaze, and you'll find a wealth of information. Firemen don't share their stories much outside the firehouse; these men aren't looking for attention or acclaim. But their stories need to be told, recorded, and passed on, because they teach us about a way of life that is noble, rare, and fast-disappearing in America.

I heard about firefighting when I was growing up because my father and my uncles were firemen. I didn't join the Fire Department, but the job still had a hold on me. A few years ago I went to see my cousin Joe at his firehouse in Brooklyn one night and, when the alarm bells rang, he practically shoved me into the front seat of the rig, right between him and the driver. We tore through the streets of Brooklyn with the siren wailing.

When, from a mile away, we were able to see the smoke, Joe banged on the glass partition to signal to the guys in the back that it was a real job. On-scene, the rig screeched to a halt and all six guys sprinted up the stairs of the involved house. Two minutes later, the structure flashed over with flames. Feeling the heat from the side-walk, I panicked, thinking of those guys inside. But then I saw one guy make it out okay and then another. A half hour later, the fire extin-guished, all six men were back in the rig. I was thinking: *What a life*.

In January 2001, I was back in Brooklyn at Rescue 2, the firehouse my Uncle Ray Downey had commanded for fourteen years. I was there for the long haul, to shoot a television documentary, and I had chosen Rescue 2 because of their reputation for being New York's busiest firefighters. Rescue 2 was where the modern rescue company was born (in the seventies) and refined (in the eighties and nineties). Over those thirty years, the house became synonymous with Rescue firefighting and pioneered many rescue methods now used world-wide. Rescue 2's reputation was born in hot, smoky hallways and in fiery collapses. If you were a fireman in trouble, the Rescue 2 men would come and save you. They were obsessed with being the best firemen they could possibly be.

Rescue 2 also became experts in every kind of emergency; if you were trapped in the City of New York—under a train, drowning in the East River, pinned in a spectacular car wreck, or buried in a building collapse—these were the men with the tools and the knowledge to save your life. They trained for every eventuality—nuclear bombs, chemical weapons, helicopter crashes. Under Ray Downey, Rescue 2 made a science out of rescue work. He helped to organize the rescues into the Special Operations Command (SOC) and created a rescue school for the next generation of rescue firemen. His men responded not just to disasters in the City of New York, but also went all across the country to work for the Federal Emergency Management Agency as an Urban Search and Rescue team. After the first World Trade Center bombing and the Oklahoma City bombing, in the mid-nineties, the rescues added new equipment and training to respond to urban terrorism. They continued to do their traditional work of saving civilians and firemen, but also took on these new and frightening tasks.

FOR MONTHS I practically lived at the firehouse. I ate pork sausages and cavatelli at the kitchen table with the guys, shared a lot of laughs and, like any new guy in the firehouse, served as the butt of many jokes. I got to know the guys well, spending night shifts cruising the borough with them. There was Lincoln Quappe, a fierce redhead perpetually gnawing a cigar. And John Napolitano, who used to stay up into the wee hours cramming for the lieutenant's exam. I can still see John sprinting down the street, into the fires, faster than any of the old-timers. I grew particularly fond of Kevin O'Rourke, the nicest guy in the Rescue, who always inquired after my mom and dad. The men came in all shapes, sizes, and demeanors. But all the men loved their job. And they all loved Rescue 2.

I started to love the place too. I even started to think about becoming a fireman. I followed them to burning buildings and saw the men run out of air, and then vomit or just collapse on the sidewalk. I had microphones and cameras with them inside the fire, so I could hear the calm before it got bad, the urgent shouting when it got really hot, and the elation after they knocked down a blaze.

I began to write down some of the stories the guys told me when I realized that a lot of the Rescue 2 experience could only be captured in a book. I wanted to write about the war years of the sixties and seventies, when the ghettoes of New York City were burning day and night, and about the pain the company felt when they lost a fireman in the line of duty. I also wanted to write about some of the incredible rescues that these guys pulled off.

By 2001, when I was there, New York's rescue companies were at the top of their game. They had battle-hardened men in key positions and they had honed their skills at countless fires and emergencies over the years. The men had been to all the biggest infernos in New York and to mammoth disasters like the Oklahoma City bombing, but none of that prepared them for what happened in September.

As I was just starting to work on the book, 9/11 changed every New York fireman's life and seriously disabled the rescue companies. In this book, I want to show what the city and the country lost when these special forces, which had risen so far, suffered a devastating setback. For most of the story I tell, I return to the years before 9/11, to what seems like a long, long time ago. Those years were good years for the men of Rescue 2. They were innocent years, when a fully involved apartment building was a thrill and no one thought much about suicide bombings or Islamic terrorists; all that seemed unreal compared to the fires burning just around the corner from the firehouse. These were the years that formed the men who responded on 9/11.

This book is about only a small group of men within one fire company. I am writing, wherever possible, about the men I know best. There are many other men who have labored as long and as hard as the men in these pages, have risked as much and accomplished as much, but go unrecognized here. There are other men at Rescue 2 and men at every company in the City of New York who are as brave, as strong, and as heroic as the men in this book. A century ago a New York fire chief was asked who was the bravest fireman. "Who can say bravest?" he replied. "We are all the brave." He is still right: all firemen are the brave. This book is about a few of those men and the firehouse they built.

THE LINGO may confuse the uninitiated. Here are a few explanatory notes:

When you report a fire in New York, there are four kinds of companies that may respond: engines, trucks (also known as ladders), squads, and rescues. Engines carry the hoses and hook up to the hydrants to put water on the blaze. Trucks search for victims and vent the blaze by smashing windows and cutting open the roof. Squads, like rescues, are special units that respond only to bigger fires and emergencies.

Rescues ride on an enormous truck filled with special equipment that no other companies carry. The guys call their truck a big toolbox, and it's filled to the brim with everything from chemical weapons antidotes to the jaws of life. Rescue companies do ordinary firefighting plus all kinds of special operations like high-angle rope rescue, scuba, and confined space operations (somebody caught in a shaft or other small space).

A typical fire goes like this: The truckies get in first and start to search the place for victims, and also knock out any windows. One of the truckies goes to the roof and cuts a hole or opens the stairwell door for ventilation, while another guy climbs the fire escape and smashes windows and looks for victims. Meanwhile the engine company unfurls hose and hooks up to a hydrant. A chief pulls up in a Chevy Suburban, driven by his aide. He stations himself out front, where he can get a look at the fire and direct everyone on scene. The arriving companies report in to the chief, who runs the show.

New York's five rescues, seven squads, Marine Division (fireboats), and Hazardous Materials Company are organized under the banner of the Special Operations Command Battalion. The SOC battalion operates as an independent entity in the FDNY, with its own training school (at the Rock on Randalls Island), headquarters (on Roosevelt Island), and chain of command (rescues report to SOC, not to their local battalion chiefs).

Each company has about twenty-five men on its roster and, at any given time, six of these men are at work. Officially, firemen work a sequence of day tours and night tours that add up to about forty-eight hours a week. In practice, the vast majority of firemen swap their tours with one another to make twenty-four-hour shifts; after working a twenty-four, they have about three days off. Setting this up is called "making a mutual" and the tour itself is known as a mutual or a twenty-four, though the two parts of the tour are still known as the day tour (9 AM–6 PM) and the night tour (6 PM–9 AM).

An alarm is called a box, a word that comes from the ancient pneumatic pull boxes that used to line the streets. Every intersection in the city has a box number, a numeric code that identifies that place to firemen and fire dispatchers. Even if there hasn't been an actual box at the location for a century, it's still called a box. When a box comes in, the fire dispatcher will send out units to respond. The closest units to the location of the alarm are "first-due" to the box and they get what is considered to be the choice assignment; they go right into the flames. "Second-due" and "third-due" units come in as backup and usually fan out to the rest of the building or to the buildings next door to the fire, to make sure the fire hasn't spread.

There are two kinds of radios in use: the handy-talkies that firemen carry into a blaze (used for communication at the fire) and the borough-wide radios in the cabs of the fire trucks and with the chiefs (used to communicate with the borough dispatcher). The handy-talkies can only be heard for a few blocks around the fire; the borough-wide radios can be heard throughout the city.

Firemen often speak in code when they talk on the radio, with numbers flying back and forth. The dispatcher is known simply as "Brooklyn" or "Manhattan," depending on the borough, as in "Brooklyn calling Rescue 2." Ten codes are shorthand for a variety of situations. Most common are: 10-75 (a structural fire that requires a full response of three engines, two trucks, a battalion chief, a squad, and a rescue), 10-8 (in the rig and available for duty), 10-92 (false alarm), 10-45 (fire victim found), and 10-60 (major emergency).

# ONE: THE JOB

FIREFIGHTING HAS ITS OWN LANGUAGE, cadences, dreams. For Bob Galione, a sixteen-year veteran of the New York City Fire Department, the words, rhythms, and hopes of firefighting define life. His job isn't what he does; it's who he is and how he lives.

On this icy blue morning, February 5, 1996, Bob prepares for work at Rescue 2, his Brooklyn firehouse, after breakfast on the run with his three girls and his wife, Linda. He's still tired from his last shift; he can never catch up on sleep. The cracks around his eyes, and the creases encircling his mouth, make him look older than his thirty-eight years. Bob's round head is nobly dusted with gray, and a bushy mustache snakes across his upper lip. Worn big and virtually unkempt, it looks like a tropical caterpillar crawling across his face. Before air tanks and breathing devices the mustache worked as a ready filter for soot and smoke. Now it's just a fashion accessory, a part of the uniform that never comes off.

Bob's home is in quiet Ronkonkoma, a Long Island suburb sixty miles east of the city in Suffolk County. Suffolk was sparsely populated farmland until the sixties, when droves of white, working-class families transformed its old potato farms into subdivisions. Bob moved here from Brooklyn years later, in the eighties, for his kids. He couldn't afford the safe parts of the borough anymore. Over the course of Bob's lifetime, everything in Brooklyn has changed. Once upon a time, if a guy said he was from Flatbush or Crown Heights, Bob could tell if he was Irish or Italian, rich or poor, a fighter or someone who'd leave him in the lurch. But now people and places change fast; today's Pakistani cabby hangout is tomorrow's Starbucks.

Bob faces a two-hour commute to his firehouse, but who's complaining? For him, the FDNY is not a job, it's a passion that began when he first jumped on a fire truck, siren roaring and lights flashing. Now, having jumped on thousands of times more, Bob is a little older, a lot more experienced, but no less enthusiastic than he was on his first run. To him, firefighting is always simply "the job" and when he utters those words, they take on a life of their own. The job lives and breathes in the thousands of men who congregate in firehouses all over the city. Firemen live in an old-fashioned world of black and white. Cops might agonize over right and wrong, but the fireman's moral universe is simpler: Evil is a red devil that wants destruction and death. Good is a charged hose line, full of water to fight the flames. A fireman down—trapped, burned, or dead—is the ultimate evil.

Rescue 2 and the four other companies like it (one per New York City borough) are expert, elite squads who get the biggest blazes and the toughest emergencies, but their special mission is saving fellow firemen and others in danger. Sometimes that means just yelling, "This way out" to a lost man. Sometimes it means getting right up inside the heart of a fire, where no human is meant to be, only to discover a brother burned to death. Once in a while, when things go just right, Bob goes deep into the heat and emerges dragging a rookie who's out of air, a little toasted but still breathing. At those moments, walking through a wall of flames makes perfect sense.

In the eyes of the city fathers, rescue men are the same as regular firefighters and get no extra pay or perks. Bob could probably live off his salary in Sioux Falls or Akron, but in New York, where humble homes sixty miles from the city can fetch a third of a million dollars, it's more like lunch money, to be supplemented by a second source of income and his wife's wages. So why stay? If forced to justify himself, he takes the party line: the job is a noble calling, a helping profession. Truth is, it's also tremendous, exuberant fun—a way to make a living and never get bored. Bob braves the flames because the few things that beat the rush and beauty of firefighting—the birth of his kids, fine summer nights with the family, waking up next to the woman he loves—aren't things a working-class guy from Brooklyn can do for a living.

Bob wants fires to burn tonight. His dream isn't a good night's sleep uninterrupted by alarm bells; it's a night packed with fires. Not that he wants people's houses to burn. No way. But if a building happens to light up, he wants Rescue 2 to be the first firemen in and the last men out. Last December, 1995, Bob worked one fifteen-hour tour with four real fires. Back at the firehouse between jobs his eyes were swollen from smoke, too painful to close. Next morning, after two more fires, half blinded by the smoke and soot, he staggered into the kitchen, feeling for the freezer. An ice cube on each eye brought down the swelling. Later, at home, he went straight for a Price Club–sized jar of painkillers and popped six. His smoke headache still lasted two days.

A big fire brings all kinds of pain: not only smoke headaches, but also sore joints, a perpetually runny nose, deep cuts and bruises from crawling in the dark, lungs and throat filled with black mucus, and a powerful nausea that makes him bend over and retch.

A good job kicks Bob's ass; it's three floors of fire, fifteen rooms going. But to Bob, a good job is worth anything. When he stumbles out of a good fire he feels like he's just fought a force of nature and won. No matter how banged up, burned up, and bloodied he is, he feels like he can conquer the world. A bad job makes him feel small, powerless, insignificant; it's where people, especially firemen or kids, get killed.

LATER THAT DAY, at precisely two o'clock, Bob bounds toward the door, carrying his wallet, keys, and, most importantly, his cigarettes. Though the night shift starts at six in the evening, he leaves himself four hours to get to work, settle in, and gab with the departing day shift. Bob's a talker.

A few minutes after two, Bob pulls his Mitsubishi four by four onto a county road littered with prefab strip malls, ubiquitous chain stores, and fly-by-night franchises. Traffic on the Southern State Parkway shouldn't be bad, though he's seen the road turn into a parking lot at three in the morning. His journeys back and forth to Brooklyn are the only peace and quiet Bob enjoys all day. Shuttling between a home with three young girls and a firehouse with five other guys, he finds it hard to tell which is louder. Others might balk at spending two hours on the road to work, but Bob just leans back in his seat, opens a pack of Marlboro reds, and inhales. He blows smoke out the window like he's exhaling smoke from a Cuban cigar.

The radio weatherman calls for snow, and that may mean work. Winter is fire season. Cheap electric heaters ignite tattered curtains. Squatters burn down slums. But on a cold winter's day with no fires, the penned-up guys are like stir-crazy kids. Instead of "Ma, why can't we go outside?" it's "Why is it so motherfucking slow around here?" In the past few years Bob has seen many days like this. Too many. Fires have decreased dramatically since the sixties and seventies—the war years when Rescue 2 regularly went to eight or ten fires a night. Airborne bottles and "fuck you's" greeted the mostly white firemen storming the projects.

Now, in 1996, the economy is up, and fires are down; that's the way it works. You can leave your car outside the Rescue 2 firehouse in Bedford-Stuyvesant without the stereo getting ripped off. People are renovating their apartments, not torching them. Guys like Bob are bored. But firemen are still getting killed. The latest wave of fatalities started with the Watts Street Fire in 1994 when Captain John Drennan and two of his men were trapped behind a jammed metal door during a superheated flashover. The gases around them heated up, and everything burst into flames. By the time other firemen sprang the door open, one of the men was dead and the other two were horribly burned. Drennan spent forty days in the burn center and then, a day after doctors made their first, cautious prediction that he might survive, he died.

Since then, seven more firemen have died. Ten men buried between 1994 and 1996. It once took a decade to kill ten firemen.

Wives keep asking their husbands, "When are you going to retire?"
But that's the kind of question Bob tends not to hear—it's impossible
to run into burning buildings if he's always thinking about getting
killed.

From Bob's vantage point, the danger seems real but remote. It's
something that happens to other people, not to him or his friends—
and definitely not to anyone in Rescue 2, where no one has died in a
fire since 1957. Even that death was a fluke, a heart attack outside the
blaze, not a death in the flames. In the same way that kids can't imagine growing old, Bob can't imagine Rescue 2 losing a man. He figures
he's more likely to go down in an ugly wreck on the parkway than die
in a fire, though he maintains a carefully controlled speed of sixty-
eight miles per hour. (Cops only write speeding tickets for violations
over seventy.)

There's no "Welcome to New York City" sign, but Bob realizes
he's crossed the border when the buildings get bigger, the air dirtier,
the drivers crazier. Passing Exit 24 on the Grand Central Parkway and
crossing into Queens, he reaches for the power switch on his Radio
Shack scanner. The Fire Department bandwidth will tell him what
he's missing. He's buffing: listening to the department radio when he's
not working, just for kicks. Some civilian buffs spend their weekends
speeding around the city, scanners in hand, trying to catch a glimpse
of the firemen at work. Each borough has a frequency of its own,
except for the eternally fireless Staten Island, which shares one with
the Bronx. Bob tunes in to Brooklyn, and the first static crackle tells
him that Rescue 2 is heading out:

*"Three-ten to Brooklyn, let's give a 10-75 to start with."* Engine
310 is calling Brooklyn with a job.

*Nice and casual,* Bob thinks, *10-75 to start with.* As if there might
be something bigger to come. A 10-75, however, is a decent blaze—a
building fire that calls for three engines, two trucks, a squad, and a
rescue company. That means that right now, deep in the heart of
Brooklyn, his buddies are heading out without him. But Bob isn't too
disappointed he's missing this job because it sounds sedate. He didn't
hear that note of panic in the officer's voice on the radio that means it's

going to be good. A big job inspires fear, even in seasoned firemen. And you can usually hear that in a guy's voice.

ERRY COYLE IS THUMBING THROUGH the *New York Post*, digesting his lunch, when he hears the call for the 10-75 over the radio in Rescue 2's kitchen. A 10-75 is a real fire—not some malfunctioning automatic alarm or false alarm—and Coyle gets a rush just from hearing the call. After eight years at Rescue 2, Coyle has worked his way up to the coveted chauffeur's spot. He has short brown hair and the rough good looks of a guy in one of those once omnipresent cigarette ads. Coyle weighs in at a muscular 175, well below the mean at Rescue 2.

Coyle arrived this morning for the day shift while Bob was still getting his girls off to school. Upstairs in the locker room, Coyle had, as usual, stowed away his clean jeans and put on dark blue, fireproof pants. His boots, coat, heavy bunker pants, and leather helmet are ready downstairs, near the rig.

Coyle loves morning time in the firehouse; the night shift is winding down, and the day shift is gearing up. Instead of the usual crew of six men regularly assigned to each tour, there are eight or ten guys hanging around and that many more stories. The *Daily News*, the *Post*, and *Newsday* (but never the *New York Times*) are always on the table, and a fresh pot of drip coffee is ready to drink. This time of day is filled with the promise of hard work and a hot meal.

This morning, as usual, the officers changed shifts at 8:45, but the firemen waited until nine sharp. The Rescue 2 roster carries a full complement of twenty-five firemen and four officers (one captain and three lieutenants). At any given time, one of these officers and five of these men are protecting Brooklyn. This morning Lieutenant Pete Lund (also known as Vulcan, as in Vulcan the fire god) led roll call and assigned positions. Lund is a short, squat man who barks orders and wisecracks out of the side of his mouth. With many years behind him in Brooklyn and a stint as an officer in the Bronx's Rescue 3, he commands respect when he goes to a fire. Which means his guys get to work where they want to: inside the blaze.

At the beginning of each shift, Lund hands out positions. This morning Billy Eisengrein got the can, the most sought after position because it will put him next to the fire. The can man carries a six-liter fire extinguisher to hold back the flames if the engine company isn't there yet with a hose. Otherwise he helps the engine move in the line. Coyle is the chauffeur, a position for a senior man. To drive the rig, you need to know the streets and be able to entertain the boss in the cab. Tommy Donnelly got the irons—an ax used to pound open doors and a Halligan tool for all-purpose prying and smashing, especially useful for wrenching open wood-frame doors. Jimmy Jaget was the roof man. Roof takes a heavy Partner saw to vent the roof so that smoke and fire can escape from the building. Louis (pronounced Lou-ie) Valentino was the hook. Hook carries a four-foot iron hook, used to rip down ceilings and get at fire hiding above.

The guys trust one another to man these positions, even the bad and boring ones. If a fireman gets distracted and doesn't cover his post, he puts everyone in danger. One of the highest compliments the men pay is, "He's always in position." That means much more than making a rescue or grabbing a hose line. It means a man can be trusted to do what's right.

Coyle steps into his boots, pulls up his suspenders, and climbs into the chauffeur's seat. He stows his jacket and air tank in a back compartment because he hates driving with his coat on. The diesel engine groans to life, signaling that Coyle is ready to pull out.

"Sounds like a job," Lund grunts, climbing into the cab while the others jump onto the back of the truck. Louis Valentino waits for the rig to clear the entry before pushing the garage door button. Then he leaps on while the truck is moving slowly, with a practiced hop, skip, and jump.

Louis' black hair and olive skin reveal his roots, stretching directly back to southern Italy. Though he's married and works a side job, Louis still hits the gym almost every day, and the contours of his muscles are visible through his bulky coat. On the Rescue 2 totem pole Louis is near the bottom; he's only been here two years. But he's taken in every bit of knowledge the guys can throw at him.

"Glenwood Road and Forty-eighth Street," Lund yells, reading off a computer printout. As soon as Coyle hears the address, he has their route mapped out in his mind. He knows where the other fire trucks will be coming from, and he'll pay special attention to the corners where their paths are likely to cross. Civilian vehicles are a nuisance, but not a real danger to the rescue truck; other FDNY rigs, however, can be deadly.

The day is brutally cold, one of those winter afternoons with a perfect cloudless sky and temperatures cold enough to freeze every drop of water on the streets. Coyle concentrates on his driving. The streets are too icy for brakes, so he downshifts at intersections, relying on Lund to stomp on the siren pedal. Its heavy bass rumble rocks the cab and keeps all but the maniacs away from the truck's twenty tons.

Coyle tries to stick to his side of the two-way street, but, if traffic is stopped, he will veer into the oncoming lane and floor it, darting around the jam. This is New York, world capital of indifference, so most drivers just stare blankly at the rig and veer to the side at the last minute. But even jaded New York City kids are not immune to the lure of the big red truck; they wave and smile as the firemen glide by.

In the back the men stand in a narrow corridor, holding on to overhead straps like the ones on a subway, putting on gear as they go. Bouncing along Utica Avenue, every man listens intently to the radio. All the extra air bottles, tools, and supplies in the back bounce against one another, producing a clanging metal symphony timed to the stops and starts of the truck.

The radio tells them that things are proceeding as usual. Right now, at Glenwood Road and Forty-eighth Street, the engine company's chauffeur will be hooking the hose to a hydrant. If the fire is put out fast, Rescue 2's race through Brooklyn will be just a joy ride. But if the engine company can't find a hydrant or the truck company can't vent the blaze, the fire will escalate. Left unchecked, a blaze doubles in intensity every thirty seconds.

As the siren wails, the brownstones of Crown Heights give way to the much larger apartment houses of East Flatbush. They're right in the center of the borough, in the Caribbean heartland of Brooklyn. Nearby are buildings filled with immigrants from Haiti, Jamaica, and

Trinidad. For the last ten years this has been the busiest fire locale in the city.

Louis Valentino stands in the back of the truck, getting his bearings. Born and raised in the borough, he knows Brooklyn well. As Utica widens, Louis hoists his forty-five-minute air cylinder onto his back. Leaning forward, he tightens the straps and pulls the bulky tank close to him so it won't get snagged on something.

The radio buzzes with exchanges between the chief on the scene and the borough fire dispatcher. *"We're using all hands at Box 2408. We've got a one-story commercial building, one hundred by forty. Exposure one is the yard, exposure two is an occupied commercial building, exposure three is the yard, and exposure four is the yard,"* says the chief, surveying the adjacent buildings that might catch on fire. Only one wall of the fire building is attached to another building. That means fewer places for it to spread.

*"Ten-four, Battalion 41. At this time you have Engine 309, 248, Battalion 33, and Rescue 2 assigned on the 10-75."*

"Sounds like something," Lund says to Coyle.

"Maybe we'll finally get some work."

*"Battalion 41 to Brooklyn."*

*"Go ahead, 41."*

*"Second alarm on Box 2408. Fire is doubtful at this time."* That means the fire is not under control.

*"Ten-four, Battalion 41."* Then two loud, long bleeps and *"A second alarm has been sounded for Box 2408, Glenwood Road and Forty-eighth Street."*

The dispatcher sends out more units—five engines, two trucks, and two chiefs—at a furious pace as the mood on the Rescue 2 truck soars. This one is for real, and the men prep for battle—checking air, tightening straps, and giving themselves the once-over.

Louis hefts a metal hook in his hand, takes a test breath from his mask. Propping his leather helmet on his head, he pulls on fireproof gloves. As the truck glides to a halt, the air brakes hiss. Someone yells, "We're in." It's been fourteen minutes since the alarm was sounded.

Out here in the industrial part of Flatlands it doesn't look like the rest of the congested borough. There's space and air around each structure, and the streets are empty of cars. This section of Brooklyn looks more like the industrial ruins of Pittsburgh or Youngstown than New York. When they step off the truck, the men see an enormous cement building that looks to be a gym, with a gated annex in the rear where the fire is burning.

Louis bounds ahead of the pack and greets the chief in charge, Fred Gallagher, Louis' former chief and Rescue 2 captain from 1973 to 1980. "Hey, Chief," Louis says. "Can you believe they made me backup chauffeur in the rescue? What's this company comin' to?"

"They picked the best guy for the job," says Gallagher, who commanded Louis until two years ago when the young firefighter made the jump to Rescue 2.

Three years ago Gallagher had a job with Louis nearby, on Rogers Avenue. After seeing the fire go bad, Gallagher ordered his men to pull out. Louis Valentino was out on the curb when that building blew up and the explosion knocked him to his knees. Gallagher picked up Louis and carried him across the street to safety.

Later, Louis' dad sent Gallagher a fruit basket. "You gave me the life of my son," Louis' dad told him.

Gallagher, who is normally a tiger, has held his men back today, because he doesn't like the look of the fire building. It's the same kind of feeling he had three years ago at that explosion. At most fires Gallagher is the one pushing guys farther in, demanding that they keep advancing the hose line and knocking down the flames. But today, years of instinct and experience are telling him, *Be careful, take it easy, hold 'em back*. He's told the engine men to first extinguish what they can from outside, before venturing in.

"Have a couple of guys check the exposure," Gallagher tells Lieutenant Lund, and Lund dispatches two men to the attached building to make sure the fire isn't spreading there.

As Louis moves toward the fire, Coyle still stands at the back of the rig, carefully strapping on his air tank and checking his mask. Then he starts to jog slowly toward the fire, following the hose lines and conserving his energy. Like every other fireman, he carries a hundred pounds of gear on his back. Following the smoke and activity, he hones in on the shack behind the health club that is spitting smoke over the nearby Long Island Railroad tracks; that's where the guys went in. Tucked out of sight, the place could have burned for hours without drawing any attention. Time means everything when fighting a fire. If a blaze is allowed to burn freely, it might cause a building collapse when the supporting beams and joists melt from the heat.

COYLE IS SUPREMELY DISAPPOINTED when he surveys what little fire remains to be fought. Wisps of smoke drift from the roof, but there are no sparks or flames.

Inside he finds the other Rescue 2 guys. Lund is near the door. "Follow Louis," he orders Coyle. "And back up the guys on the line." The place is littered with old cars. It might be a chop shop to harvest stolen car parts. Coyle follows the line up to Louis.

"Hey, Louis, what's happening?" says Coyle.

"Nothing," says Louis. "It's over."

Small pockets of fire tease the engine company men, who spray the ceilings and walls to chase down the flames. Louis and Coyle climb over car parts and walk behind the nozzle. The adrenaline has worn off, and Coyle feels only cold water dripping down from the ceiling and freezing on his coat. He wanders into a small hallway crowded with firemen, then veers off to the left and tells Louis, "I'm gonna check this room here."

Louis sticks near the hose line. Coyle hears an engine guy ask Louis to hug the wall so that they can get the stiff hose around the bend. Out of the corner of his eye, Coyle watches Louis jump out of the way.

In the next room, Coyle starts poking and prodding the piles of rags that litter the floor. He wants to make sure that there is no fire hiding in the debris. Suddenly, he hears a snap, like a wooden plank being split with an ax, then a much louder ka-boom. He dives to the floor as the roof and walls crumble around him. Firemen scream, and

Maydays go out over the radio. But nobody can answer the calls. They're all buried.

Coyle's first thought is to get air. As he hears the men around him climbing to the surface, he claws his way toward sunlight.

BOB'S RADIO SHACK SCANNER always craps out on the old Interboro Parkway (later renamed the Jackie Robinson Parkway), the small lick of road that cuts across the no-man's-land between Queens and Brooklyn. So Bob travels the gentle bends of the parkway with no idea of what Rescue 2 is experiencing. He hopes Rescue 2 is in there battling a blaze, making a push down a long fiery hallway.

The parkway comes to an end, and the traffic spills into East New York, the heart of the ghetto. After an incomprehensible syllable, Bob catches some cryptic radio transmissions that indicate someone is badly hurt at the fire. When a fireman is injured, the dispatcher communicates on a private frequency, so Bob can't make out what's really going on. He can only tell that it's bad. Must be an injured fireman from another company or a civilian.

Bob speeds up a little because he's eager to find out what happened. He shoots down Atlantic Avenue, braking at red lights but then rolling through intersections when possible. The elevated train rumbles overhead. He has worked in this neighborhood for seventeen years. He grew up in Bergen Beach but has spent his entire Fire Department career in a twelve-block radius, right on the border of Crown Heights and Bed-Stuy. Bob likes to say that this is where he learned to be a fireman, and where he learned to be a man. He knows every gambling den, drug warehouse, and illegal squat in the area, and he can tell you the tale behind every burned-out vacant building.

Bob started out in a company just down Bergen Street known as the Tin House Truck because it was quartered in a cheap tin shack that the city never got around to renovating. When Bob came to the company as a probie (probationary firefighter), the officer told him, "Hold on to my jacket wherever we go inside a fire." He liked to keep the probie close, so he didn't have to worry about this new guy getting killed. Bob was like a puppy with its mother. The officer would walk, crawl, and slither his way through the fire, and Bob would be right behind him, grasping his coat, as the officer pointed things out. "That room's too hot for us," he'd tell Bob, or "Let's check under the bed." After sticking with him and not doing anything stupid for a few months, Bob finally earned the officer's respect. When the officer told Bob to check a room by himself, Bob knew he had become a fireman.

The guys at the Tin House hated Rescue 2, which, in their eyes, got the glory without the grunt work. Rescue would come in, help put out a fire or save somebody, and then speed off to another job. Everyone else would be left standing there, exhausted, with three hundred feet of hose to help pack and fold. And, in the eyes of Tin House veterans, the worst thing you could ever do was give the jump to a rescue guy. Rescue 2 men were supposed to support the other companies, but they also loved to get the hose line and put out the fire on their own. Giving up the line to Rescue was worse than getting burned. It meant months of humiliation in the Tin House kitchen.

Snaking around a semi, Bob darts through a yellow light and hits the home stretch, the final block of Bergen before the firehouse. This is the beat of the Seventy-seventh police precinct, which for years has had the greatest number of murders in the city. On the corner stands Hannibal's, a small take-out shop that seems to scrape by with no business save an occasional Italian ice sold to a neighborhood kid. Next door is Randy's Hideaway, a tiny bar frequented by black customers that's the only beacon after night falls. Bob's four by four darts past impounded cars that the federal marshal has hauled to Bergen Street. Across from the firehouse, grammar school kids pour joyously out of school, bundled up in puffy down jackets and brightly colored winter hats.

Rescue 2's firehouse is dwarfed by the high-rise Albany housing projects one block west. The park behind the firehouse is mostly deserted now, but at night it comes to life as an underground gun market. Prospective buyers are entitled to a few test shots before plopping down their wads of cash to buy weapons. Pete Martin, a Rescue 2 fireman and a gun buff, can distinguish the makes and models of the handguns by the sounds of their shots.

Bob slams to a stop in front of the Rescue 2 parking lot and hops out to open the padlocked gate. The firehouse is ringed with twenty-foot-high fencing topped off with twisted razor wire, but even this has not been enough to deter all thieves. Many times the men have returned to find their lockers stripped of valuables and the television set gone. After a spate of break-ins the FDNY assigned an injured fireman to a twenty-four-hour-a-day security detail at Rescue 2.

As Bob pulls his car into the lot, the radio gets busy, but he still can't figure out what went wrong with the 10-75. Bob checks vehicles. Pete Lund's beat-up Chevy tells him Vulcan is the officer in charge. Next to it is Louis' convertible jeep. In the summer, Louis likes to drive with its top down all the way from his lifeguarding job in Coney Island to the firehouse in Crown Heights. Bob always tells Louis he's crazy to go through the heart of Brooklyn exposed like that. Coyle must be driving the rescue rig, because his truck is parked right next to Louis' jeep.

After hurriedly parking his car, Bob walks to the rear door of the firehouse just as he did on the first day he came here, seven years ago. He had an interview with Ray Downey then, and he was nervous—scared really. Like every other Brooklyn fireman, he knew of the captain's commanding presence, iron-clad lungs, and steely gaze. He had seen Downey bound up a tenement stairway thick with black smoke without even using a face mask for air.

Rarely had a captain so dominated his men or his house. Downey helped make Rescue 2 into one of the toughest, most efficient, and most respected firefighting institutions in New York. Gradually, the house's reputation grew beyond the city. Nationwide, other rescue men adopted Rescue 2's ghetto-tested methods, and Downey's crew grew into some of the most experienced firemen anywhere. They were a group of guys who could handle everything, and whose ego, drive, tenacity, and humor gave the house its reputation for toughness. Many of the men went on to captain their own units. Through the years, Downey's reputation continued to grow. In the winter of 1992, when Flight 405 went down into the water near LaGuardia Airport, he climbed to the top of the sinking fuselage and yelled out orders to all the rescuers on scene, cops and firemen alike.

Downey joined the FDNY at the start of the war years, in 1963. These fire-filled years comprised his first decade on the job. As a fireman in the early seventies, he watched all hell break loose and fought more fires in a day than many firemen of the nineties see in a year. Raised in Woodside, Queens (a working-class Irish stop on the Number 7 train), he had come through the marines and expected his men to follow him wordlessly wherever he led them. They always did.

When Bob first walked in those firehouse doors to face Downey, he was intimidated by Downey's reputation. But the captain didn't say much. He just stared long and hard at Bob, asked him a couple of simple questions, and then later conducted the real interview with Bob's captain at the Tin House. *Had Bob cracked under pressure and bailed out of any fires? Would he lead others into action?*

Downey telephoned Bob a few weeks later and told him that he had the next open spot at Rescue 2. Nowadays Downey has moved on to head all New York City rescue operations. But still, every time Bob enters Rescue 2, he feels a little bit of the same excitement he felt when he first walked in seven years ago, thrilled at the prospect of working for a legendary commander like Ray Downey. Today, after what he's heard on the radio, that familiar excitement is mixed with foreboding.

TERRY COYLE IS STUNNED, buried under the pile. Everything that went into building the garage is now layered on top of him. After a few seconds of silence, everyone starts shouting at once. Coyle points his head toward the sunlight, working his way through roofing, Sheetrock, and wood. Cold snow leaks into his gloved hands, and he heaves his body up out of the rubble. Taking a deep breath of fresh, cold air, he starts to climb across the pile to see who needs help.

When the Maydays stop, there is a quiet sense of relief. Most of the firemen around Coyle have been lucky. The inside of the building was littered with old cars, and almost everyone dove underneath a chassis just as the roof came down. A few firemen are working on a stunned Chief Gallagher, his face covered in blood. The collapse knocked him out, but now he's coming to.

There are enough guys helping Gallagher, so Coyle wades through the debris, scanning the pile for other firemen in trouble. A piece of shiny black rubber catches his eye. Two boots protrude from the rubble. He gets on the radio: "*Rescue chauffeur to battalion. Mayday. We have a man trapped about ten feet from the door.*"

The boots remind Coyle of the trapped witch in *The Wizard of Oz*. It's hard to believe that anyone is attached to them. Coyle starts knocking away debris around the boots, trying to see what's holding the man down. Then he finds one of the red flashlights Lund and Louis always carry. When he looks more closely at the brand of leather boots, he confirms that it has to be either Lund or Louis pinned in the collapse. Shimmying under the pile, Coyle tries to get to the man. He takes off his air tank. He hugs the ground. It's still impossible to get underneath, so he stretches out his arm and feels a limp gloved hand. Removing the glove, he grasps the wrist, praying for a pulse. But he feels nothing. Then he hears Lund's voice over his shoulder. So this must be Louis knocked down in the rubble. Coyle is burrowing under car parts, roofing, and beams to get to him. Louis has no pulse. If they don't get him out now and start CPR, he's dead. On top of the debris Lund and the others are starting to move pieces of the building off of Louis' body. They are knee-deep in junk, trying to figure out what's pinning Louis.

Coyle backs out and tells Lund, "It's Louis, boss. I can't get a pulse." Lund sends Eisengrein running to the rig for air bags and orders Jimmy Jaget and Coyle to cut around Louis to free him.

Coyle positions his hand under the saw blade and guides it around Louis' body, with his hand between the saw and Louis. They cut away roofing, metal, and some thin wooden planks. With a jolt, Coyle sees that an enormous steel I beam is lying across Louis' chest. When he grabs it and tries to lift, it doesn't budge. Lund yells for air bags again. Other firemen crowd around the tight space, trying to help. Voices rise to a crescendo. Finally, Lund screams, "Shut the fuck up so we can work here!"

The first set of air bags, from another company, won't inflate. Eventually Eisengrein arrives with Rescue 2's air bags. The men put one air bag under the beam. Coyle watches it slowly inflate, powered by an air tank, inching the beam off Louis. Coyle and Lund grasp Louis' body and hold him until he's free.

Coyle's hands shake as he pounces on Louis and starts CPR. It's hard because the beam has crushed Louis' chest. Other firefighters pass them a Stokes Basket—a caged stretcher—and Lund says, "Put Louis in the Stokes and let's get him to the hospital." They slide Louis into the stretcher as Coyle pumps his chest one last time. The men cling to the hope that the ambulance or the hospital might work some miracle and bring Louis back to life. But after ten minutes without a pulse they all know Louis' chances aren't good. Right now they can't even think about what that means. A chain of firemen pass the Stokes Basket up over the rubble toward a waiting ambulance. Coyle climbs in after Louis. He wants to ride with him to the hospital.

Inside the ambulance, two paramedics go to work. They search for a pulse, then an airway. They can't get the tube down Louis' throat because his neck is crushed. Coyle stares at his friend.

He hears the paramedic on the radio to the hospital, *"We've got a fireman with no pulse, no airway. We're three minutes out. Over."*

USUALLY BOB FEELS BETTER just being in the kitchen. It's the center of life at Rescue 2, and Bob rules it like an Italian grandmother. The most fastidious homemaker in the company, he's serious about mopping and sweeping. He spends hours in the kitchen, cooking southern Italian meals and yelling at everyone to turn the damn TV down. Though he makes an exception for *NYPD Blue*, his favorite program, he doesn't like TV at the firehouse; he likes it the old way where guys just talked.

The kitchen is empty now. The table is clean and the floor mopped, so Bob knows the guys had a slow day. On the countertop are chocolate star cookies, the kind Louis always brings. Besides Bob, the only other guy in the entire firehouse is the security detail, a guy from another company, who sits by the radio looking grim. Bob barks, "What happened?"

"It's bad," the detail says. "All of a sudden they started screaming."

Must be a fireman hurt or killed, but from what company? Then the phone rings. Bob springs up. It has to be news. Word travels fast around the Fire Department. Everybody has a cousin, brother, or friend to whom he passes on stories of triumph or tragedy. Nothing stays secret for long.

"Rescue, Fireman Galione," Bob says.

"Bob, it's Commissioner Gregory." Gregory's son, Mark, is a Rescue 2 man, so Bob figures it's a personal call, but then the commissioner's voice wavers and the hairs on Bob's neck stand up straight.

"Is there an officer around?" Gregory asks. He runs the communications branch of the FDNY. This can't be good.

"No, it's just me, boss," Bob replies. "What can I do for ya?"

"Need you to go upstairs for some phone numbers."

Bob knows what this means. The fireman in trouble, the guy he heard about on the radio, is somebody from Rescue 2. Bob can't believe it. He struggles to get out some questions. "Who? What happened?"

"I can't tell you."

No way Gregory is getting off that easy. Bob needs to know. "You have to tell me something because you need me to get those numbers," Bob says. "How bad is it?"

"As bad as it can get."

That means a brother dead or badly burned. Bob puts the phone down and climbs up the creaky stairway, thinking of the men working, remembering Louis' jeep and Lund's pickup. He knows Coyle is driving. Who is going to be the victim?

Turning left in front of a row of lockers, Bob steps into the office. On the desk is a silver box he doesn't want to look at. It hasn't been opened under these circumstances for many years. Inside are notification cards that all firemen fill out when they first come on the job: Next of kin. Blood type. Religion. All the details Bob had hoped nobody would ever need to know about a Rescue 2 guy.

He opens the box and picks up the phone extension. "Gimme a name," he tells Gregory.

"It's Louis Valentino."

Bob thumbs through to V, sees a creased index card with faded handwriting. Slowly, he reads out Diane Valentino's home phone number and address in Bensonhurst, Brooklyn. Gregory repeats it all back to Bob. Gregory's next move will be to send out Chief Ray Downey, a chaplain, and a union official to bring Diane the news.

Bob feels it right in the center of his chest. Short of breath, he's sicker than he's ever been in a fire. He wants to rip the house apart.

Fifteen minutes later, the phone rings again. This time it's Downey. As always, he gets right to the point. "Bob, I need you to check that address again."

"I got the card in front of me. You got the right place. What's the problem, chief?"

"Does she work?" asks Downey.

Bob knows that every afternoon at three o'clock Louis fights for the phone so that he can check in with Diane after she comes home from work. She has to be there.

"She works, but she'd be home by now, boss."

"Gotta go, Bob. I just saw her looking out the window."

DIANE VALENTINO ALWAYS WALKS home from the lawyer's office where she works. It's under the el, and the houses fan out from the overhead train in neat rows. The old men staring at a soccer game and sipping espressos laced with anisette in a neighborhood café remind Diane of her own childhood in Carroll Gardens, Brooklyn, another Italian enclave. She and Louis have always hoped to move back there someday.

After getting married, Louis and Diane moved out to Staten Island to save money for a place in Brooklyn, but Louis missed everything about his borough, from the Brooklyn-Queens Expressway (BQE) to his favorite eggplant, fried peppers, and provolone sandwich at Defonte's. So, after a couple of years saving, Diane and Louis came back home.

When Diane thinks of her husband's job, she visualizes the men joking around or preparing a meal. The danger is something she's always tried not to think about. There is, after all, nothing she can do about it. If she started focusing on the risks Louis takes, she'd drive herself and her husband crazy. Louis has explained to her what he does and made it seem routine, not dangerous. Sometimes when they go out to dinner, he spots a job and trails the engine to the fire, following a block behind. He shows Diane everything: the engine hooking up to the hydrant, the truckies ripping open the roof. His confidence reassures her. And his boundless enthusiasm about fire-fighting, even when he's just watching from the sidelines, convinces Diane that there is no other job for Louis, no matter how much this one worries her.

Lately it's been hard not to think of the risks. Every couple of weeks, it seems, Louis is sending his dress uniform out to the cleaners, polishing his shoes, and setting out on his day off for a funeral in Queens or Long Island. Diane has comforted herself with the thought that Rescue 2 is different. Louis always told her that rescue was his insurance policy. The men were trained to save trapped firemen, and if he ever got stuck, they'd pull him out. She had almost believed him.

Outside the Valentinos' house, the temperature has dipped down to sixteen degrees. When the doorbell rings, Diane wonders who is wandering around on a day like today. Her curtains are closed for privacy, so she walks over to the bay window and peeks out. She sees three firemen, all dressed up. Recognizing Ray Downey's shock of white hair, she wonders what he's doing in Bensonhurst.

RIGHT NOW BOB WANTS the brothers to fill the firehouse. He wants to hear a joke or the guys laughing after a joke. But he hears nothing. And he can't tell anyone the news. Diane's and Louis' parents deserve to know first. So maybe it's better no one else from Rescue 2 is around.

Word about the collapse starts to circulate in Brooklyn firehouses as the companies from Glenwood Road file back home. Soon everyone working in Brooklyn suspects that they will have to salute another fallen fireman. They just don't know who. When off-duty Rescue 2 guys start to hear that something went wrong, they call the firehouse.

Bob fields their calls. All he can say is, "I don't know what happened, but why don't you head in?"

Jay Fischler, the current captain of Rescue 2, arrives. Gregory must have called Fischler at home, Bob figures. Fischler nods a greeting to Bob and rushes up to his office to take care of details.

Fischler replaced Downey in late 1994. Downey has been a tough act to follow, and Fischler, like every captain before him, has had to go through a rough breaking-in period. Rescue 2 is full of personalities—big, loud personalities who have a lot to say about everything. Fischler must feel like hell with a Rescue 2 man dead.

Soon the men scheduled to work start to trickle in. Bob doesn't let on that he knows anything. He just hears the speculation and waits. It seems like he waits forever. When the Rescue 2 rig turns the corner onto Bergen, Bob hears the fan belt start to whir. He walks out of the kitchen and onto the apparatus floor, where the truck gets parked, waiting for the men to back the rig into the firehouse. Waiting to hear the news.

Bob presses a button, and the creaking and clanging begin as the old wooden door slowly rises up to let the truck in. The other men trail Bob out of the kitchen, and they stand by as the firemen on duty climb down from the truck. The youngest men, Tommy Donnelly and Jimmy Jaget, look stunned.

Each of the men waiting in the house finds his closest friends to talk to. Bob navigates around small groups of firemen to the far side of the rig, where Coyle sits in the driver's seat not moving or speaking, as if all talk has been sucked out of him. Finally, he cuts the ignition and hops down from the cab.

Seeing the look on Coyle's face, Bob loses all hope of Louis being alive. Coyle's eyes are full of pain.

He says to Bob, "You heard?"

"I think I know what happened."

"Louis is dead," says Coyle.

In the past, talk has taken the sting out of anything. If Bob had seen a civilian get cooked, he would throw that out in the kitchen. Just mentioning it would help. But right now Bob doesn't think talk is going to help anyone. Louis is dead. What more can you say?

Still, Bob wants details: what killed Louis, why couldn't he be saved, how painful was it for him? He needs to know what happened. Coyle is badly shaken and injured, but telling the story steadies him like a shot of whiskey.

As he listens to Coyle talk about waiting in the hospital for news about Louis, Bob can't believe what he's hearing. Louis was laid out on a gurney, Coyle says. The men circled around his body, and there were quick prayers and farewells. They touched Louis' arm or grasped his hand. After a little while Lund ordered them out to the truck, and they headed back to the firehouse after the worst run of their lives.

Bob doesn't know how to fit this into the scheme of things. All these years he has told his wife that he had the best guys on the job backing him up. But he also knew that Rescue 2 went to the most dangerous fires, and that it was one of the riskiest jobs in the department. Only two guys had died in the company's seventy years. And, until today, no Rescue 2 man had ever gone down in a fire.

Bob tells Coyle, "It was a freak thing." That was it. A freak thing. Not a normal part of the job. What hit Louis was pure bad luck, like lightning striking a man down. What happened was the kind of tragedy that could get anyone, not just a fireman.

Some of the guys don't agree. Dave Van Vorst, a wise old veteran, says, "I just can't believe it doesn't happen more often." But the consensus remains: it's an exception, not something that can happen again, not something that can happen to any one of them. It wasn't something about the job. It was a freak thing. And that becomes their mantra: a freak thing. Day after day, month after month, they'll call it that.

Ray Downey drops Diane off at her mom's house in Carroll Gardens before proceeding to Louis' parents' house. Louis had always loved to bring the Rescue 2 firemen to visit his mom and dad in Red Hook, Brooklyn. If they had a fire nearby, he would insist they stop for a visit. Louis senior was a longshoreman who had worked his way up to New York State assistant labor commissioner. He has the kind of grace that is only found in those who have made their own way in the world.

Louis' mom and dad greet Downey at their doorstep. Diane was shocked, mute, but the Valentinos burst into tears immediately. For hours, while Downey makes plans for the funeral and organizes the pipe and drums corps, the church, and the funeral home, the old couple cry for their son. Relatives stream in from the neighborhood and then begin to arrive from farther away.

When Louis' best friend, Frank Campesi, another firefighter, arrives, Downey tells him, "You're assigned to this house. Don't go to your firehouse for work. Come here." The Valentinos used to joke that Frank could pass for Louis' brother, and although Frank is a little heavier, and Louis was a little darker, they shared the same bodybuilder's figure and wise-guy expression. Frank and the family don't say a word for two hours; they just suffer in silence, unable to stop crying.

Frank thinks about the last time he saw Louis at a fireman's funeral in Queens. "Jesus Christ, Frank," Louis had said. "What the hell is goin' on with all these funerals? It could be any one of us."

Louis and Frank first met in the sweaty, subterranean Fifteenth Street Gym, in Park Slope, Brooklyn, a place with pictures of local boxers plastered to the walls and no yuppie Stairmasters or ab bars. The gym had only the ancient free weights that serious bodybuilders favored. Talking one day, Louis and Frank realized they were both preparing for the Fire Department physical, a grueling obstacle course. Together they rented an abandoned airport hangar at Floyd Bennett Field, outfitting it just like the real test. Every weekend they trained there. That was twelve years ago.

Now Frank sits down with Chief Downey and the Valentinos at their large wooden dining-room table, and Downey tells them, "We'll take care of everything. Let's just figure out what you want for Louis."

AT FIRST THE MEN AT RESCUE 2 just hug and cry, and Bob is no different. As the tears stream down his face, he sees the photo on the kitchen wall of Louis, Coyle, and himself getting tattooed with the Rescue 2 bulldog, the company emblem. Their smiling faces contrast with their bruised biceps. Louis had been reluctant to tell his dad about getting the tattoo. When the guys at Rescue 2 found this out, they roasted him mercilessly in the kitchen. "Big tough Italian guy, and you can't even tell your daddy you got a tattoo. What's this place coming to when we let in people like you?" Now as they all learn of Louis' death, everyone migrates into the kitchen for a cup of coffee, and the men start telling stories like this one—funny stories that absorb a little bit of their pain.

Coyle thinks of another Louis story. Influenced by his summers lifeguarding at Coney Island, Louis had worn three little pooka beads around his neck. That was the height of fashion in Venice Beach, but not in Bensonhurst, where a gold chain was much more popular. One day the guys had each put three tubes of ziti on strings around their necks and waited for Louis to notice. He looked from one neck to the next and cracked a smile, even as he protested, "I'm starting not to like you guys."

Then there was the way the guys persecuted Louis for his refined Italian palate. One big Irishman, Mike Quinn, would insist on ordering bologna on white bread when they went into the best *salumerias* in Brooklyn. Back at the firehouse Mike would always douse his pasta with ketchup, just to get Louis riled up.

The laughter feels good, as if Louis is still alive and with them in the kitchen. As the men recall every Louis story they can remember, Bob thinks about going back into a fire. It would feel good tonight if they could catch a few jobs. They need to get right back into it. So he talks to each man who is slated to work. They decide to get the tools in order and clean up the rig in preparation for another run.

Then the doorbell rings. Bob walks out to answer the door, and he sees a chief's car blocking in the Rescue 2 rig. When the chief notices Bob staring at his vehicle, he tells Bob, "You're out of service."

Later that night a throng of reporters stands outside of the firehouse, calling out questions to the men when they walk out the door. The guys even catch a photographer who snuck inside taking pictures of photos on the firehouse wall, trying to find one of Louis.

The next day, when they open the morning newspaper, the men see a photo of Louis taken from a picture on the wall. In the newspaper photo he's smoking a cigarette. Louis didn't smoke, but he's immortalized with a little graffiti cigarette in his mouth that someone had drawn on the firehouse photo last year.

For ten days Rescue 2 remains out of commission. The guys come to the firehouse every day and gather in the kitchen. They run errands for Diane and the Valentinos. They make sure that every last detail of the funeral is taken care of. When they wake up early in the morning, Lieutenant Al Fuentes tells a couple of guys to head over to the funeral home and sit with Louis. He doesn't want him to be alone in the morning, when everyone else is waking up together.

FOR THE FUNERAL Louis returns to the place where he was baptized, made his first communion, and was married: the Church of Sacred Heart and Saint Stephen in Carroll Gardens. Its Gothic spire is perched next to the BQE and when the pipe and drums start playing they can't completely drown out the sound of traffic heading into the city. Storeowners shutter their shops and dress up for mass. Small, ancient Italian men huddle on street corners in their Sunday best, talking about losing one of their own. It reminds some of the older men of the war years gone by—World War II, Korea, and Vietnam—when they would salute a local hero this way every few months.

The firemen who knew Louis best pack the church, along with his relatives and friends. Thousands of other firemen linger outside in the cold, rubbing together their gloved hands and sending the young guys around the corner for hot coffees. The funeral is a ritual familiar to most of these men and one of the only occasions where the whole department comes together. Inside the church they pray and sing; outside they talk, some reminiscing about Louis, others making plans to get together after the funeral. Some religions condense mourning into one concentrated burst of grief—the service and burial. But the men of the FDNY, following the traditions of New York Irish and Italian Catholicism, make every death into a social occasion, where people don't just cry but also laugh and talk.

Ten thousand firemen line the streets of Brooklyn and salute, as one, to send Louis off to heaven. Nearest to Louis are the men of Rescue 2. Bob and Coyle help carry the coffin. The rest of the Rescue 2 men march slowly behind them. Many past members of Rescue 2 are also lined up on the steps of the church. Pete Bondy, now retired, a legend from the seventies to the nineties, still looks young in his uniform jacket. He's grown his salt-and-pepper mustache down to his chin. Fred Gallagher, chief at the fire where Louis died, stands saluting the caisson. Al Fuentes, Louis' lieutenant, is one of the men who eulogize him. "Louis was a shining star," he says.

All of the men of Rescue 2, past, present, and future, are assembled here. Out in the crowd are the men who have been dreaming of joining this company ever since they came on the job. Men like Lincoln Quappe, a fireman from a truck company in Crown Heights who professes to hate Rescue 2 but who would come in a heartbeat if he could. And Phil Ruvolo, a lieutenant, soon to be captain, who is hoping to get a rescue company to command. Today these men are not thinking of their ambition; they are thinking only of Louis Valentino.

TWO DAYS AFTER THE FUNERAL, Rescue 2 goes back into service. There is nothing better than a fire to prove that the job can still be done. Each time Bob drives back to the firehouse he sees Louis' jeep in the parking lot. It makes him think of what he doesn't want to be thinking about when he's going into work. But if Bob starts to believe that Louis' death might have implications for his own life, he just repeats the mantra: a freak thing. It was even what he told his wife, Linda, when they drove home from the funeral.

In front of the brothers, doubts can never show. Civilians might think that what propels firemen into that last dangerous room to search for a victim is an extraordinary heroism. That's part of it. But Bob knows that the other part of the equation is peer pressure. Even now, when Bob walks into a burning building, he won't flinch. If he does, they'll eat him alive. Can't ever back down. Can't ever retreat. Not with the brothers watching his every move.

# TWO:

## THE NEW CAPTAIN

Not long ago, Phil Ruvolo was just one of the many firemen lost in a massive sea of dark, uniform blue outside Louis Valentino's funeral mass. Now Ruvolo has become an FDNY captain. He's looked forward to this day since he first became a fireman. Ruvolo is getting his own firehouse today, and not just any firehouse. He's going to be the commander of Rescue 2.

It's been cold, but on this February morning in 1998, his first day of work at Rescue 2, spring is peeking out of winter. As he pulls his old SUV onto the sidewalk, next to the firehouse in Crown Heights, he studies Rescue 2's bright red, automated garage door, decorated with a mural in memory of Louis Valentino.

Ruvolo knew Louis well. They had worked together for two years at Ladder 147. When Louis came to the ladder company, Ruvolo was a senior man and Louis was a Johnny, a new guy. Ruvolo knows that Louis' death has left a huge mark on Rescue 2, a house that had prided itself on never losing a man. Two years after his death, Rescue 2 is still in mourning. But Louis' death is only one of the things Ruvolo will have to deal with at Rescue 2. He knows that even before they've met him, a good portion of the guys inside already hate his guts. And not just because he's here to boss them around.

Ruvolo has a history with some of Rescue 2's veterans, a pack of men who know how to nurse a grudge so that it lasts forever. Back in

1992, when he was a lieutenant at Staten Island's Rescue 5, Ruvolo successfully lobbied to have some Brooklyn alarm boxes assigned to his company—which meant that they were taken away from Rescue 2. Ruvolo argued that the boxes, located in Coney Island, just across the Verrazano Bridge, were closer to his Staten Island Rescue firehouse than to Rescue 2. As far as the senior men in Rescue 2 were concerned, it was infringement on their sacred territory—and it would be remembered as long as Phil Ruvolo dared to set foot on a fire truck in the City of New York. Who was Ruvolo to mess with seventy years of tradition just because his company didn't see enough action in its own borough?

Six years have passed, but Ruvolo has it on good authority that Richie Evers, the senior man at Rescue 2, still gets so angry every time he hears the radio dispatcher calling Rescue 5 to one of Rescue 2's boxes that his blood pressure skyrockets and he looks as though the veins in his temples are ready to burst. Ruvolo, without much prompting, could come up with a dozen more stories like this, tales that explain just what it is that makes the Rescue 2 men such legendary pains in the ass. But the guys savor this reputation. They just love to piss people off. And, along with fighting fires, that's one of the things they're best at.

As a newly assigned captain, Ruvolo doesn't have a key to Rescue 2's front door, so he has to ring the bell. As he presses his finger to the buzzer, he feels a surge of excitement and a jolt of anxiety. Ruvolo takes a little breath of fresh air before he enters the maelstrom. Probably Richie is in the kitchen right now cursing him out. Ruvolo will play it cool, but he's a little relieved when a young fireman answers the door—a new guy whom Ruvolo doesn't even know. Nodding good morning, the kid holds the front door open for Rescue 2's next captain.

Rescue 2's men have always enjoyed the brief periods when they are without an officially assigned captain. That time is a kind of vacation from authority. The house's three lieutenants earn respect, but they do not equal a captain. As one of Ruvolo's old captains put it, "The difference between lieutenant and captain is like the difference between your mom and dad. Mom can always say, 'Wait till your father comes home.'" Ruvolo is like a father coming home after three years away on business. Who knows if Mom or the kids are in charge?

RUVOLO GOT THE CAPTAIN'S SPOT, a highly coveted position, through a combination of hard work, talent, and good luck. The job of rescue captain is one that the commissioner and the chief of department are definitely consulted about; they need to sign off on who is going to occupy such a vital position in their city. Ruvolo knew Fire Commissioner Thomas Von Essen, and he was able to do him a favor a few months before he got the job at Rescue 2. At that time Ruvolo got a call from Von Essen, telling him to meet one of Mayor Giuliani's aides at the midtown Hilton. It was all pretty mysterious. Dress casual, no uniform, Von Essen had said, and do *whatever* the mayor wants. Ruvolo was known as the FDNY's rope specialist. He was a great firefighter and a skilled scuba diver, but his special expertise was ropes. When the day came and Ruvolo walked in, the mayor's aide identified him immediately. Even in khakis and a polo shirt Ruvolo looked like a fireman. "You the rope guy?" asked the aide. "I need you to make the mayor fly."

Mayor Rudolph Giuliani was organizing his annual roast for the New York press corps. The roast was Giuliani's chance to get back at reporters and torment them, the same way they had tormented him all year. Traditionally, Giuliani donned stockings and a girdle and did a song-and-dance number in drag. That year, he wanted to be airborne on stage—a sort of Peter Pan effect. So Ruvolo rigged an elaborate set of pulleys, ropes, and harnesses, hoisted Rudy sixty feet off the ground, and swung him across the stage. He made the mayor fly and earned a powerful ally in the process. The mayor seemed to like Ruvolo, a no-nonsense guy who could strap him into a harness and make him fly through the air without making a big deal of it.

Ruvolo saw Giuliani again at Randall's Island for the FDNY promotion ceremony where he became a captain. After shaking his hand, Giuliani gave Ruvolo and all the other new captains two small silver bars that pinned onto the collar of their shirts. But passing the exam and getting the bars were just the beginning of getting a *good* captain's spot. Ruvolo never saw himself at the helm of just any house. There are firehouses busy night and day, and firehouses where you get bored watching *The Price Is Right* and taking afternoon naps. Ruvolo wanted to be where the action is: Rescue 2. So for the last four months, he's campaigned for the job and renewed old friendships with people who could help. Ruvolo has called in every favor he accumulated from his years of fire service. He has worked hard and done the grunt work to earn the spot—not just the little things like making the mayor fly, but the big things, like running the SOC's rescue school and paying dues as a lieutenant in Staten Island's Rescue 5. He feels incredibly lucky to get the Rescue 2 job. With only five rescues in the city, some guys have waited their entire careers for a captain's position in a rescue house. Many have had to settle for something else. For Ruvolo, it has all worked out—his hard work and his experience have convinced the people in charge that he is the man to lead Rescue 2 into the future.

THE GUYS in the kitchen figure they're ready for Ruvolo. They give every new captain visited on them by headquarters a hard time, and they're more than ready for a new target. They plan to pick their teeth with Ruvolo's captain's bars after they've consumed his tender carcass. Rescue 2's symbol is a snarling bulldog with its teeth bared like a wolf, and the guys do their best to live up to the image. The captains who have been able to control them had to rule with iron fists. Ray Downey was known to verbally eviscerate people so badly that they would think twice about even speaking when he was around. His mere presence led one firefighter, Timmy Stackpole, to revert to a childhood stutter that hadn't been heard for twenty years. Downey's predecessor, Captain Fred Gallagher, had an equally persuasive non-verbal approach. If a fireman fucked up, he would put him in a head-lock. During Gallagher's seven-year reign, forty-two firefighters departed from Rescue 2, and Gallagher couldn't have cared less; if they couldn't put up with him, they didn't deserve to be in Rescue 2. It wasn't for everyone.

Considering this checkered past, a new captain might be forgiven for avoiding the kitchen—the heart of the beast—during the first moments of his maiden voyage into Rescue 2. But not Phil Ruvolo. Here is a man determined to let the company know he has arrived, and that he's in charge from this second on. He blows right by the guy holding the front door and heads straight for battle. He walks up to the kitchen door as if he's going to kick it in.

Ruvolo is built like a football player who has relaxed a little bit in the off season; his body retains the broad contours of fitness but not every muscular detail. With his prominent nose and jutting, harshly sculpted chin he could be a stand-in for Robert De Niro. His black hair is cropped tight in a flattop, and a few wisps of gray peek out. Where men like Bob Galione function at a modulated pace, like engines perpetually warmed up, Ruvolo operates with a manic intensity, bouncing from delight to rage in a matter of seconds. When he walks into a room, he takes it over, throwing out knowing glances and knuckle-busting handshakes and making the place his own. He has a dark sense of humor that can squeeze a laugh out of even the bleakest moment. When he aims a barb at someone, he doesn't stop until he's hit the bull's-eye. Then he bursts out with a maniacal guffaw that draws everyone in, even his victim.

This morning, the formidable Richie Evers sits at the center of the kitchen crowd. A tough, old-school fireman, Richie stands six-foot-three, with the requisite mustache and a booming voice that fills a whole fire floor and, at times, a good part of the state. He's been at Rescue 2 almost twenty years. When he first came into the company, the hazing was brutal. Early in his Rescue 2 career, he got his hand stuck in the soda machine. A senior Rescue 2 fireman came up to him and asked, "Are you sure you can't wiggle it out?"

"No," Richie said. "It's stuck." The other fireman punched him in the face and walked away, knowing that Richie wouldn't be able to come after him.

The hazing he endured helped make Richie the proudest, most vigilant protector of Rescue 2's honor. Once, when an officer from another company was covering for a night at Rescue 2, the substitute took a huge feed of smoke and leaned over to vomit. Storming after him, Richie screamed, "Go back to the rig and do that! Nobody sees us puke."

Officers come and go. But it is men like Richie who are the bedrock of Rescue 2. He has an encyclopedic knowledge of all the tools and he can run drills when the captain has to be upstairs in his office doing paperwork. But more important than the technical knowledge Richie will pass on to new recruits is the way that he embodies the spirit of the firehouse. He always goes at a fire with the same determined intensity. Richie never goes easy on anyone else; but he also never goes easy on himself. Though his rank might be the same as thousands of firemen in the department, this rank means nothing: he's a leader, whatever his rank is. And because he commands so much respect in the firehouse, Richie can be an officer's best friend, or his worst nightmare.

Richie—unbent, unbroken, and definitely unimpressed—stares down Ruvolo, who just saunters over to the coffee machine and picks up the pot like an Indian Chaiwallah, lifting it high into the air and arcing the liquid down for a three-foot drop into his plastic coffee cup. Then he nods a greeting to Richie, getting a fuck-you stare in return. But Ruvolo also notices some friendlier faces, especially among the new guys. Lincoln Quappe, a redhead who Ruvolo knows just recently joined the company, gives the captain a smile. Apparently, he hasn't yet learned Rescue 2's welcoming attitude or become privy to the fact that Ruvolo is guilty of a crime against humanity, or at least against Rescue 2. The only guy who dares to actually say hello is the other officer, Lieutenant Pete Lund, Vulcan the fire god.

After he downs the cup of coffee, Ruvolo gets the key to the gated parking lot, goes back outside, and pulls his truck off the street. Then he decides to check out the captain's office. Dropping his fire gear off near the front of the rig, he climbs the stairs, carrying his uniform pants and a freshly laundered, powder-blue officer's work shirt.

The captain's tiny office beckons at the head of the stairs, next to the locker room, and Ruvolo, all by himself for a moment in the quiet room, feels a quick swell of pride as he sits down at the same ancient wooden desk where Fred Gallagher and Ray Downey once worked. The battered old relic looks like a holdover from when Rescue 2 opened in 1925, and it takes up a quarter of Ruvolo's new living space.

Against the right-hand wall, away from the window, is a twin bed, with a metal frame and no box spring. A half step up from an army cot. No wonder Downey was always so cranky.

Glancing around the room, Ruvolo sees that every available inch of counter space is covered in files, papers, or forms. He sits down and digs into a mound of paperwork, starting his daily slog through the bureaucracy, slaving away with ancient tools to copy, sort, and collate a baffling array of forms. The FDNY has promised computers to its officers, but for now things are still done the old way, on paper, with only the occasional use of a modern contraption such as the type-writer.

By the time his shift begins, Ruvolo has changed into his uniform. He heads downstairs for roll call. Richie is chauffeur. Ruvolo doles out the rest of the positions. He isn't planning on doing anything drastic just yet. First, he needs to prove himself inside a few fires with these men, and then show his stuff in the kitchen, where, after fires, the men go to recount the action, swap stories, and use what has just tran-spired in fires to train new members. The men will be watching Ruvolo like hawks, just waiting for him to lose his temper or his cool. He has to show them that he is worthy of leading them. And he has to do it fast.

THERE IS HARDLY a day when Ruvolo regrets his decision to become a fireman, even though every firehouse has its complications and he often has to put up with enough psychodrama to fill a month's worth of after-school movies. Ruvolo grew up in Flatlands, Brooklyn, but at age fourteen, in 1969, he started trekking into Manhattan to be educated by the Jesuits at Saint Francis Xavier High School. Although he'd never admit it in front of the firemen, he spent a year abroad in Paris and learned French. Later, after he finished college, he sat down to size up the paths before him: one was the fireman's life of his father and grandfather and the other was a more conventional, professional path. He knew that a job in accounting at a place like Brooklyn Union Gas would pay him much more than he'd get for risking his life as a fireman. But he didn't care. He knew he was a fireman—despite the Jesuits' French.

Then, as now, there was intense competition for the city's limited number of firefighting jobs. Ruvolo passed the test with flying colors but spent a couple of years doing odd jobs and waiting for a free spot in a firehouse. During this seemingly endless wait, he spent a lot of hours talking to his dad about the job. His dad told Phil a lot about what he would encounter when he came on the job. But there were some things he left out. Years later Ruvolo learned that his own grandfather, John O'Connor, had died from injuries suffered while fighting a fire in the early forties. There was no high-tech treatment for burns back then, and Ruvolo's grandfather perished from a secondary infection, weeks after the fire. The whole time he was growing up, Ruvolo's parents had never said a word about how his grandfather died. That was the part of the job no one in a firefighting family ever wanted to talk about.

In 1978, a couple of months before Ruvolo joined the FDNY, his cousin Bill O'Connor joined the department. Ruvolo was envious that Bill, his best friend and the man whom he had chosen to be best man at his wedding, got to join before he did. With just a month on the job, Bill was called to a fire at a Waldbaum's supermarket right near Ruvolo's neighborhood. Bill climbed up to the roof and started to cut a wide hole to vent the smoke and gases. All of a sudden the building collapsed, throwing Bill and five other firemen into the flames. All six died, making it one of the worst days in FDNY history and definitely one of the worst days in Ruvolo's life. Yet even after burying his best friend, Ruvolo came on the FDNY with the gung ho enthusiasm of youth. Somehow he felt that nothing would touch him. He doesn't feel all that immortal anymore, especially after going to forty-eight firefighter funerals in the past nineteen years. But he's determined to lead this house and make all the suffering he's been through add up to something good.

A few minutes after he goes back up to his office, there's a quiet knock on the door. He yells, "C'mon in," and Lincoln Quappe slinks in.

Before Lincoln came to Rescue 2, Bob had summed him up for the other guys in the kitchen. "He's that redheaded guy from 123 who's always mother-fucking Rescue 2. The guy that wouldn't piss on us if we were on fire." But Lincoln has made a rapid transformation from one of the biggest Rescue 2 haters in the borough of Brooklyn to a disciple of Richie Evers, the house's most faithful son.

When Lincoln first arrived, Bob wouldn't let anybody else answer the front door. He opened it like a butler and had a shit-eating grin on his face as he waved the new guy in. Bob didn't even have to say a word. Lincoln knew just what he was in for and realized that his previous antirescue rhetoric had definitely been noted. But now, just a few months later, Lincoln is like a born-again Christian who has just surfaced from his immersion baptism. He's repented all his rescue-hating sins, and Rescue 2 has become the way, the truth, and the light.

Lincoln throws a furtive glance out to the locker room, to make sure no one has seen him enter, and then says, "Look, Captain, don't be offended, but I'm not gonna be able to talk to you when Richie's around." Ruvolo just chuckles and watches Lincoln sneak back out.

When the first alarm of the day sounds, Phil Ruvolo sprints down the stairs, taking them two at a time, hoping it's a good job. But when someone in the back hands him the printout, Ruvolo sees that it's nothing. There are a few nearby boxes that Rescue 2 covers, and they tend to bring routine fires like this one. Most of the time, for these minor jobs, a rescue captain just stands back and lets the other engine and truck companies do the work.

When Rescue 2 pulls up to the apartment building, Ruvolo heads into the fire building and down a flight of stairs to investigate. The scene is typical for these fire-starved times: about twenty firemen are on hand to extinguish a smoldering oil burner that probably would have put itself out. They drown it in gallons of water and the floor is flooded with slick, burned oil. The outing's most dramatic moment comes when Ruvolo, stepping up to jump in the rig, slips on the oil on his boots and lands flat on his ass. Dusting himself off, he looks up to see a big smile on Richie Evers's face.

But that snicker is all Ruvolo hears out of Richie all day. Richie won't say a word to him. When it's time to eat lunch, Richie gets on the intercom to the guys in the back and says, *"Would someone ask the captain where he wants to go for lunch?"*

*I love it,* thinks Ruvolo. *The guy won't even talk to me, but he's letting me choose our lunch spot.* Ruvolo preempts the broadcast from the back and transmits, *"Tell Richie anywhere."*

Firemen hunt down the best lunch spots in the city. For most companies this search is confined to the fifteen or twenty blocks around their firehouse—their response area. But for Rescue 2, with the run of Brooklyn, a whole borough's worth of pizza joints, sandwich shops, and cheap Chinese takeout is at their disposal. Their rating system is a little different from Zagats, though. An enormous hero at a low price beats a gourmet *ciabatta* in a refined setting. Portion size is everything. All their regular spots know to throw on a few bonus slices of roast beef or stuff in an extra meatball for their firefighting clientele. Today, Ruvolo sits up front and watches the rig while Richie and the other guys go inside the deli. Then they head back to the firehouse to eat the meal.

Richie has probably backed this fire truck into this firehouse ten thousand times. But today, just as he's gliding comfortably in, he catches a rearview mirror on the side of the door, and it breaks right off. Ruvolo's mouth is filled with hot pastrami, but he doesn't let that stop him from doing his best imitation of Richie's snicker. *Touché.*

RUVOLO'S FIRST FEW WEEKS at Rescue 2 are pure, fire-filled fun, despite the moments of utter hostility, rebellion, and merciless hazing that make up the process of testing out the new captain.

Ruvolo has his hands full with Lincoln, who loves to start fires down on the apparatus floor. He likes to fill a metal trash can with paper and then light it up and let it burn. Sometimes he builds miniature cardboard cities, soaks them in kerosene, and torches them. Ruvolo runs downstairs one day to find the apparatus floor filled with smoke. "One more fire down here and you're out," he tells Lincoln. But Lincoln will find another way to make trouble and have fun. One of his other tricks is what he calls "stirring the pot." He gets a good argument started in the kitchen and then walks out in the middle, leaving the other guys screaming. It's like starting a fire, only with people. Even back in 1989, during the Crown Heights riots, when firemen were barricaded into their houses, Lincoln found a way to have fun. He and another fireman climbed to the roof of their firehouse, then leaped onto a neighboring building and worked their way down the block on the rooftops, performing an urban reconnaissance mission for their own amusement. They watched the looters hit a store on the corner and then saw an army of police storm the area.

Lincoln always wears shorts and a mock turtleneck at work, his own take on the uniform. In the mornings he is absolutely reliable: the first thing he does is clean out the fridge and shovel whatever is left over from last night's meal right down his throat. He can eat five big bags of potato chips before dinner and then pack away a full meal, but he still stays thinner than the other guys. Like many of the best Rescue 2 firemen, Lincoln is what the guys call a "character." That word is always uttered with a slight inflection—a muffled chuckle—as if the guy saying it is, right then, laughing a little at the exploits of the character.

For firemen who want action, Rescue 2 are the kings of Brooklyn. Brooklyn is vast, with a population so large it would be the fourth largest city in the United States if it broke away from the rest of New York City. The place has it all: the faded glory of seaside Coney Island, the Russian quarter of Brighton Beach, the Bushwick barrio, and pricey penthouses that rise along the promenade in Brooklyn Heights. Brooklyn has its own accents, favorite foods (pizza, Nathan's hot dogs, and Italian-style roast beef heroes), and a stunning rainbow of semi-peaceful, if occasionally irritable, ethnic groups. Haitians have staked their claim to a large swath of central Brooklyn, the Flatbush area. White yuppies have conquered Park Slope, on the west side of Prospect Park. And Bedford-Stuyvesant remains one of the oldest and strongest African-American neighborhoods in the city—a working-man's version of Harlem. With its thriving diversity and combination of residential, commercial, and industrial sections, Brooklyn offers everything a fireman wants: quick-moving house fires with residents to pull out, pitched battles in giant factories, and all-night jobs in neighborhoods filled with neglected, vacant buildings just waiting to burn.

Four days after he takes over Rescue 2, Ruvolo pulls up to a fire in a three-story railroad-style apartment building in Williamsburg. He and Eddie "Pit Bull" Rall share a hose line inside. At times the fire is so intense that they have to stop spraying water forward, and shoot over their backs. Later, as they move to the second floor of the building, Steve Brown, the rescue roof man radios down to Ruvolo, "Cap, you're on borrowed time. You better make a move." The roof is going to come down on them if they don't put the fire out fast. So Ruvolo and Eddie each take a line, backed up by Bill Esposito and Joe Jardin, and race upstairs to the top floor. After extinguishing the fire, they walk out steaming hot but satisfied.

Ruvolo and the Pit Bull bond in that chaos. After a good job like this, the men traditionally retire to the kitchen to brag, kvetch, and talk for hours. Some of them don't ever seem to leave. They even sleep on the kitchen couches overnight.

Ruvolo knows the ways of the firehouse kitchen from his years on the job. But during his first weeks in Crown Heights, he gets to know the particulars of Rescue 2's fabled meeting room. There are some basic, unspoken rules governing after-fire conversations: No one ever talks about the fear; admitting this sort of weakness would be considered an infraction of the Rescue 2 code. If Ruvolo is scared shitless, he might say, using the popular technique of extreme understatement, "It was about time for us to go." Likewise, Rescue 2 men never concede that a fire was hot. Ruvolo will say of a really scorching fire, "It got a little warm in there." And the men, of course, will understand. New recruits quickly learn that understatement is part of the game.

Not long after the Williamsburg fire, Ruvolo sits at the round kitchen table, dressed in his bunker gear, chatting with the men. Dave Van Vorst, an experienced senior man with a taste for practical jokes, comes into the kitchen, and there is a sense that something is about to happen. First he sprays Ruvolo's leg with some mysterious substance and then runs out the door squirting a trail of liquid behind him. A few seconds later, Ruvolo notices a flame winding along the liquid trail and quickly bearing down on his leg. It's burning like a fuse in a Road Runner cartoon, except much faster. Ruvolo's bunker pants erupt in flames, but there's no way he's going to give Dave the pleasure of even the slightest reaction. He's in no danger, because the gear will protect him; still he does get a little warm. But he doesn't even flinch. He just takes the cup of coffee he's drinking and pours it on his pants, dousing the flames without missing a beat. He doesn't even interrupt the conversation.

Taking a practical joke and turning it to his advantage is a way for Ruvolo to earn points with the boys in the kitchen. If he had screamed or panicked, he would have gone way down in their estimation. But just putting out the fire and ignoring it shows how he will lead them— he is one of them and, like them, unflappable.

Even though he's moved up to captain, Ruvolo still appreciates the fun of the job. He knows he will always savor moments like this, which will become part of firehouse lore and be laughed at forever. His own early years on the job were ecstatic, boisterous fun, mixed with healthy doses of adrenaline and fear, and he will never forget them.

THE FIRST BIG FIRE of Ruvolo's career was at an abandoned bar in Flatbush in the fall of 1978. He crawled in behind the nozzle man, and they walked into a true inferno. Fire raged above them, rolling across the ceiling, and dashing out from the rear of the bar into the front, where they had entered. Ruvolo held the line and dug his heels into the floor. He braced his shoulder against the nozzle man to keep him in position. Then he heard a voice from the bar.

Alarmed, he looked up to see one of his fellow engine men seated on a bar stool and heard another fireman tell him, "Hey, there's nothing I can do. *There's no ice.*" These two clowns were waiting for the hose to move in and pretending to order drinks from the bar. They weren't needed until the line moved in further, so they chose to make a joke out of it, and have some fun.

At that moment, Ruvolo realized something that has held true at every firehouse he's worked at: he was working with people who were borderline insane. The fire was rolling across the ceiling, and they acted like they just didn't care. Then Ruvolo felt a boot in his back. The guys in the truck company had to get past the engine to search for victims in the fire, so there was always some jostling. The truckies were kicking Ruvolo, trying to get by him, mumbling, "Come on, kid. Move it along." It didn't take long before Ruvolo was one of the guys doing the kicking.

The jobs that Ruvolo remembers from his early days are the really great fires. The ones he can never forget are the horribly bad ones. These stay with him, and with every fireman, forever. Although they are never spoken of, images, sounds, and faces remain in a fireman's consciousness for life.

Ruvolo can still hear the shouts of three children in a Flatbush apartment. The kids were behind a door. He could hear them screaming but couldn't get to them. For a minute and a half, Ruvolo and his company struggled to get the door open, but it was steel and set in concrete, and it just wasn't coming off, no matter what they did. The flames were rolling over the top of the kids, and their screams kept coming. Nothing went right. Finally, when the fire started to peek out the door, Ruvolo realized that the screams had stopped and that people could die, just like that, for a reason as stupid as the fact that a refrigerator was jammed behind their door, blocking them in. After that job, Phil Ruvolo was no Johnny anymore; he understood the weary looks he sometimes saw on the faces of veteran firemen even after a good job where everybody got out. He understood that he would carry that fire, and the children's screams, into every job he ever went to.

Just a few days after his first child, Victoria, was born, Ruvolo went to another fire that still haunts him. When his company arrived, a woman was screaming hysterically, "My baby's on the top floor." Ruvolo, his captain Mike Burke, and fireman Bill Johnson promptly climbed up the ladder attached to their fire truck, and jumped off into a window on the fourth floor. The fire was in the hallway, and the apartment door was holding it back, keeping the three men safe. Then the roof collapsed and took out the door.

Fire was chasing Ruvolo as fast as he could crawl. He realized that he had to bail out of from the fourth-floor window. As he gathered his strength to do what he knew he had to, he prayed that the ladder was still there.

But by the time he had edged close enough to see out the window, the ladder was gone. Staying calm and trying to keep his wits about him, he sighted a tree with some upper branches that he hoped would cushion his fifty-foot fall. Just then the ladder swung toward him. Maybe he could make it. No choice. If he stuck around, he'd burn to death. Nobody was coming to his rescue. So he jumped as far as he could, spreading out his arms and legs, hoping to catch a piece of the ladder. Slamming into one rung, he grabbed with his gloved hand for another. His leg wrapped around the side, and he hugged the ladder. The other two men followed. Glancing up, Ruvolo saw blue and orange flames shooting out the window he had just jumped out of.

By the time he made it down to the street, he saw the woman who had screamed for her baby. It turned out that the baby had been down on the street all the time; the mother had just panicked. Now the baby was in her arms. That was a reality lesson, a mortality lesson. He could die doing this. And, with a child of his own, it meant even more to stick around.

Phil Ruvolo realized that this was all very real, and that his kids and his buddies' kids, and a lot of people in Brooklyn depended on him. A few nights later, he was back at work, diving into the flames with abandon, and what he'd been through hadn't made him immune to a thrill or a joke. But he got it now. When he was at that window, praying for a ladder, he had understood the trust he was placing in the hands of his brothers.

BOB GALIONE STILL HOLDS COURT around the kitchen table, his booming voice often leading discussions about everything from the latest fires to the lousy politicians leading this country down the tubes. Cigarettes and seventeen years of sucking down smoke have given Bob the gravelly voice of authority—an asset around the firehouse. Smoking seems a negligible health hazard since he inhales mountains of toxic fumes at a fire. On scene, every fireman is equipped with a self-contained breathing apparatus that draws clean air from a tank on his back. But in time-honored tradition Bob saves this air until he urgently needs it and sucks smoke the rest of the time. He hopes something quick and painless will get him before asbestosis does.

Even with a stalwart like Bob running things in the kitchen, the place hasn't felt the same since Louis Valentino died. For Bob, time has moved slowly since Louis' death. At first he thought of Louis not just every day but every hour, sometimes every minute. Each man in the house was having these same thoughts, though, like Bob, they remained silent about them—and about the fact that Rescue 2 had changed. It had lost a man.

When men work together as the Rescue 2 guys do, when they have this much fun together, when they risk their lives and love every minute of it, when they yell and scream and knock down doors, when they drag one another out by the scruff of the neck from superheated infernos, they get to be like family. Everything else that civilians associate with the fireman's job—all the mushy, sentimental stuff they layer on top of this experience—is just the bonus. For Bob, it's all about the thrill of fighting fires and the bond with the men. When the bond is broken by the death of a brother fireman, it's an open wound.

The jeep that Louis once drove to his lifeguarding job stayed in the parking lot, next to the firehouse, for all the men to see, long after he passed away. At some point, in those early months after the accident, Bob pulled into the lot with a funny story on his mind, spotted the jeep, and thought *Wait'll I tell Louis. Louis's gonna laugh at this one.* But then Bob realized, *Wait a minute, Louis's not gonna laugh at anything.*

A few months after Louis' death, Coyle decided it was time for the jeep to leave the firehouse. Diane had been too heartbroken to get rid of it, and so, as time passed, it had remained in the lot, crushing the men every time they caught a glimpse of it. Coyle, who had just moved into a new house on Long Island, told the guys that he would keep Louis' jeep in his garage forever. He felt a pang of guilt about taking the jeep away from the firehouse. It was a terrible day when he drove it off. But he knew it was the right thing to do. Things had to change.

The collapse that killed Louis also injured Terry Coyle. At first, immediately after the collapse, maybe because he was so preoccupied with Louis' death, Coyle hadn't even felt his own injuries. But gradually he noticed that his hands tingled with pain. Sometimes he couldn't even hold a soda can steady enough to drink.

When the roof had caved in, a piece of it had come down on Coyle's head. The force of this blow, combined with the effects of many others like it over the years, had crushed Coyle's neck vertebrae together. This injury was so common among firemen that the Rescue 2 guys even had a name for it: snapneck. The pain wound down from Coyle's neck all the way to his fingertips. For months he suffered silently. How could he complain about his neck and arm pain when Louis had died in the same collapse?

With Louis' jeep gone from the parking lot, Bob felt better. The men still talked about Louis all the time. In their memories, he lived on in the eternal present. Everyone else had changed. Bob's knees started to go bad, and he put on some weight since he couldn't jog much anymore. Coyle's neck injury got worse, and he thought about retiring. Richie Evers started to feel the symptoms of snapneck catching up with him after years of ceilings, bricks, and cinder blocks falling on his head. But in all these men's minds Louis remained the same: a muscular, handsome thirty-eight-year-old who wouldn't age another day.

Bob often remembered the first time he met Louis—on the beach in Coney Island where Bob was hanging out with another Rescue 2 guy just before Bob was going to enter the company. Bob and Louis exchanged words. "I'm the next guy going to Rescue 2," Louis bragged to Bob.

"Cuz," said Bob, using one of his favorite salutations, "I don't think you're the next guy going to 2. I'm the next guy."

They met again, a couple of years later, when Louis finally made it to the company.

Bob was well aware of just how much Louis' death had affected everyone: Frank Campesi, Louis' best friend, had never banked on such heartache. Frank didn't look forward to going to work anymore. He didn't want to be there. He took a job driving a chief. He didn't feel the need to go into any more fires. Frank talked to Diane Valentino almost every day. Her biggest regret was that she and Louis had waited to have children. "When you're thirty-four years old, you think you have plenty of time," she told Frank.

Louis' death hit Lieutenant Pete Lund hard. He was the boss that day, and it was his job to bring all five guys back home to the firehouse. He knew Louis' death wasn't his fault; but the job would never be the same again for him. Lund's teenage son had loved to come to the firehouse with his dad. When a call came in, he'd hop on the back of the truck with the guys. Once they arrived he'd clamber out on the sidewalk and watch his dad lead the men into the fire. But now Lund's son couldn't bear to come back. He'd loved Louis too.

Louis Valentino was just one man, but his loss in 1996 brought a pain to Rescue 2 that wouldn't go away. God knows they tried to make it go away. They had memorials, funerals, and fund-raisers. They put flowers on his grave; they had his parents over for dinner. They named the new rig "Da Banger" after Louis. They had a T-shirt printed with a shining star and a lifeguard's chair on the back. But nothing could change the fact that he was gone. Louis still came into their thoughts all the time: when they wanted him there and when they didn't want him there. Louis was everywhere: in the firehouse, at fires, at the dinner table, at home. They saw Diane in the faces of their own wives and wondered how their wives would live with something like this. But the men also learned something from Louis. They learned how to live with death. They learned how to bury a man properly. And they learned that there are some things that can't be buried. They saw that their wounds would never heal completely. And as much as they didn't want to, they had to think that this could happen to any one of them.

For Bob, the period after Louis' death was about just getting through each day and trying to get the feeling back. For a long time, he was just going through the motions. The same gray abandoned houses kept lighting up. The streets at night were still filled with sirens and flashing lights when the men went out the firehouse door. Bob felt the same knot in his stomach he had always felt when he faced this force of nature. But after he had put a fire out, he didn't feel all the good things he used to feel. And that was a serious problem because the guilt, anger, and sadness that had built up in the house since Louis died might get a guy killed. It wasn't only being reckless that was dangerous. Being reticent, overly conservative, or downright fearful could get you killed on this job too. Fires didn't hold back, hesitate, or feel anything. Firemen did.

Not long after the anniversary of Louis' death, Bob got a job that made him feel good again. It was one of those events that, in retrospect, worked out perfectly. On a July night in 1997, at 2 AM, Bob found himself in the back of the rig speeding to a collapse in Flatbush. It was the night before he was taking his family to Busch Gardens for a well-deserved vacation and he really needed some shut eye, He was grumpy in the back because at first he thought the call was bogus—a false alarm. Then he heard a radio transmission from the first company on scene: *"Three-oh-nine to Brooklyn, 10-60. There's been a collapse."* A 10-60 was the FDNY code for a major emergency. Bob expected to hear the Brooklyn dispatcher's comforting reply, but instead he heard an enormous rumble on the radio. *"We just had two five-story apartment buildings come down,"* yelled the officer on scene.

Bob sprinted up to the third floor, where an old man was trapped. The smoky scene was chaos. Half of the building was down, and the rest might follow. Bob did a tightrope walk across an exposed I beam, then managed to crawl into the apartment, free the terrified old man, and drag him out over the narrow I beam to safety.

A man walked away that night who wouldn't have without Rescue 2. And Bob remembered how it felt to save a life. The old feeling was starting to return. When Ruvolo took over as captain, Bob felt another burst of possibility. Maybe the guy was no Ray Downey, but he was no slouch either.

THE DAY TOUR AT RESCUE 2 routinely begins just after 9 AM with roll call behind the rig. After that, the men spend most mornings drilling, cleaning up tools, and organizing things. Yet, all that time, they are listening attentively, waiting for a run to come in and break the place wide open. All their make-work chores will be thrown aside immediately if there is even a whisper of possibility on the borough-wide frequency or the police scanner. Strictly speaking, there is no need for the men to monitor the radio as vigilantly as they do: when a real run comes in, they can't help but hear the loud siren it triggers. But there are radios all over the firehouse: in the kitchen, on the apparatus floor, outside in the parking lot, even down in the basement. These guys like to be extra vigilant. It's their nature. If they hear something come in that might be good, they'll go for a ride toward the alarm box, just to get a jump on it, even if they are not officially assigned. They call these stealth runs. The idea is to get a little closer to where the action is just in case they're summoned.

Around eleven or twelve they'll head out to pick up lunch, often grabbing sandwiches, or, if the day is slow, buying groceries to cook. Usually, they eat around one or two, then catch a few jobs in the afternoon. By four, unless there's a job in progress, they've drifted back to the firehouse. Typically, to save on commuting time, Rescue 2 firemen combine a day and a night tour for a stretch of twenty-four hours straight, usually from 6 PM to 6 PM, which means that the change of tours in the evening is the most social time of the day. Most of the night shifters have arrived by then, and men who have to get to a child's baseball game or a second job can head out a little early as long as they get someone to cover for them.

Yesterday was one of those typical Rescue 2 days: some drilling, a couple of jobs, some laughter around the kitchen table. Today, however, Phil Ruvolo is on the way to a job that sounds like something to break up a series of days he can't distinguish from one another.

A cold breeze blows into the cab as they tear through the streets. Captain Phil Ruvolo rocks to his right, and his head taps the passenger-side window as the Rescue 2 rig careens around Grand Army Plaza, on the way to a fire in Flatbush. He's riding shotgun to Bob. Ruvolo's had a few fires with his new guys, but nothing that's told him whether Bob and the rest of the kitchen crowd are as good at fighting fires as they are at talking about them.

It's the tail end of winter, fading into spring. The ice and snow are thawing, and Ruvolo hears the rig washing through the watery streets as they approach Prospect Park. Ruvolo watches Bob cradle the enormous wheel. When Bob hits an especially sharp bend, he hunches forward to throw his body into the turn.

As they near the burning house, Ruvolo smells smoke. It's enough to make him turn and say, "Game on, Bob."

Ruvolo is out of the rig and down the block before Bob has even made it to the rear of the fire truck. The captain is not a guy to waste a second. As Ruvolo approaches the blaze, he sees flames blowing out the second-floor windows and two hoses stretched through the front door. The enormous wood house is perched on a small parcel of land with a couple of ancient trees in front, near the street. The oscillating fire-truck lamps cast patterns of light and shadow on its facade. Quickly, Ruvolo adds up the facts in his head: from the flashlights he sees in the windows, he knows that the engine and truck companies have moved in and are scattered all over the house. But they probably haven't gotten to the attic. Ruvolo will hustle up there to look for victims.

The burning house is a Queen Anne: a large, single-family Victorian-style dwelling with a standard design that usually has only slight modifications. Though this Flatbush neighborhood is now strictly working-class, Queen Annes once housed Brooklyn's landed gentry and speak to Ruvolo of another place and time. One of the reasons he loves being back in Brooklyn is that he knows every one of these standard-type Brooklyn houses better than his own home. Crawling around these houses for nineteen years, blinded by smoke, he's memorized them by feel, and this knowledge is vital. If he is suffocating in a Queen Anne after exhausting his air supply, he knows where to find a window. If a fireman gets lost on the second floor of a Queen Anne, Ruvolo knows where he is likely to be. This can't be taught in the academy. Probies can study blueprints and pass tests, but when Ruvolo gets lost in a maze filled with smoke and flames, it's the years crawling through similar mazes that get him out. Each FDNY veteran is a repository of the special wisdom and practical experience he has acquired during his time on the job. Ruvolo knows houses.

And Ruvolo loves Queen Anne fires. They get going fast, take forever to put out, and pose some special challenges. Inside a Queen Anne there are lots of hidden nooks where victims might be trapped.

There are huge rooms to search. And there's plenty of timber to burn if the fire gets roaring.

The chief, who commands from the front lawn, waves Ruvolo inside. In the hallway two hose lines branch apart. One bends sharply up the stairs, the other into the first-floor dining room. Ruvolo figures that the first-due companies have probably searched the other floors already, so he will stick to his plan to head for the dangerous attic, where there might be people trapped. The house is sparsely furnished, and firemen crowd the interior. Ruvolo squeezes around three engine men and climbs upstairs beside Lincoln Quappe, whom Ruvolo almost fails to recognize without his trademark cigar. From the stairs, the second floor looks dark. Flickers of flame and beams of a white flashlight pierce the smoke.

"More line!" yells the nozzle man.

Ruvolo starts to taste smoke—the nasty, carcinogenic kind that makes you cough and sputter for air. In years past, Downey and other veteran firefighters operated without air tanks, but the smoke they were breathing back then often came from organic materials like wood or cotton and was cleaner to breathe. Now Ruvolo is choked by the sickly smell of heavy plastic. When he finally makes it up to the second-floor hallway, he kneels down and removes his helmet so that he can slide his mask over his head. All the men in Rescue 2 wear forty-five-minute tanks as opposed to the thirty-minute tanks that ordinary companies carry. The larger tanks mean more time inside the fire but also more weight. Ruvolo considers himself lucky if a forty-five-minute bottle lasts twenty-five minutes in the extreme physical exertion of a fire. One of the reasons he tries hard to stay in shape is so that he can breathe slowly and use less air when he's fighting a fire.

Ruvolo motions to Lincoln that he's going upstairs to the third-floor attic. Lincoln, the company's can man, lingers on the second floor to back up the men from the engine company. It's dangerous for Ruvolo to go above the blaze because fire moves up a house, reaching toward the unlimited air supply on the roof. If the fire creeps toward the attic, Lincoln will warn the captain. But Ruvolo knows that if he doesn't search the attic, anyone stranded there will die from smoke inhalation.

Ruvolo climbs the stairs to the attic. If the second-floor fire spreads to the stairs he is scaling, there's no easy way out. Ruvolo could jump three stories and hope for the best, but that's risky with a hundred pounds of gear on his back. Six years ago, a lieutenant in Rescue 4, Tommy Williams, died jumping out of a third-story window. His gear weighted him down, and he landed headfirst on the concrete below.

So when he finally makes it up to the attic, Ruvolo listens closely to the radio. He's ready to race downstairs if the fire approaches. But he's proud of himself for making it up here first. Everyone else must be busy below.

Smoke obscures everything beyond two inches in front of his face, so Ruvolo navigates entirely by feel. Once he orients himself, he starts to search every inch of the floor for a body. He stays low, where he's likely to find a victim, and where the air is cooler. Starting his search pattern off the wall to his right, Ruvolo crawls a few feet, hugging the wall, and then fans out toward the center of the room, scraping his belly, his legs, and his arms across as much floor surface as he can. He looks like he's doing the breaststroke without water, probing outward with his arms and legs. After moving away from the wall, he returns to it and heads a few more feet clockwise, beginning the same search pattern from there. From years of experience he knows what a victim feels like, what a sofa feels like, what a dead dog feels like. If anyone is up here he has a good chance of finding them. As far as he knows, he's the only fireman in this attic.

Continuing to work this same search pattern, Ruvolo wonders if he'll find a bathroom installed up here. Bathrooms and bedrooms are where he usually finds victims—often suffocated in their beds or passed out in the tub. Some people think they can escape from the fire by jumping in the bathtub and turning on the water. Ruvolo finds some boxes, some furniture, but nothing that feels like a body.

Then, as he hits the halfway point around the room, his left leg makes contact with something, maybe someone. Only it doesn't feel like a victim. A victim would be limp, passed out from smoke. This feels like a linebacker. *Shit,* he thinks, *who got up here so fast?*

"Who's that?" asks the other fireman, obviously as miffed as Ruvolo to discover that he has company. Ruvolo recognizes Bob Galione's voice.

"The captain," Ruvolo replies sternly.

"Which captain," Bob inquires.

"*Your* captain, Bob," says Ruvolo, no longer able to hold back his laughter.

They check the rest of the room but find no one, then they drop down to the second floor. By this time, the fire is out, so they drag their expired air tanks and exhausted bodies back to the rig.

With a moment to rest and collect himself, Ruvolo thinks about what happened up there. Bob moved fast, almost as fast as he did, and he went for the attic too. *Okay,* thinks Ruvolo, a little reluctantly, *I've got a player here.* Bob certainly talked like a hardened ghetto fireman; now the captain knows he's got good instincts too.

Bob is roaring with laughter on the ride back to the house. He loved it when Ruvolo answered, "*Your* captain, Bob." And Ruvolo knows that Bob isn't reluctant to share a story: within minutes of the company's return, every man in the firehouse will have heard the tale of his brush with the captain in the attic. Ruvolo also understands that a healthy rivalry is brewing. Right now the score is Ruvolo one, Bob zero, because the captain made it to the attic first. But he can be sure that at the next fire Bob will be eager to even the score. These rivalries are good for firefighting. Men are able to push themselves beyond pain and exhaustion when their pride is at stake.

Later that night, in the kitchen, after Ruvolo goes up to bed, Bob passes on the word. "There's a new sheriff in town. This captain knows what he's doing," he says. "He made the attic before I did." Not too bad for a traitor from Staten Island.

A few weeks after that fire the paperwork goes through for Ruvolo's official transfer to Rescue 2. Before it had been tentative, revocable, but now it's for good, issued on a department order that goes out to every firehouse in the city. Now Ruvolo isn't just the covering captain; he's the captain, period. He starts to answer the phone with what will become his trademark line. When the guys answer a call for the captain and then pass him the line, he always says the same thing: "Captain Ruvolo"—pause—"The Rescue." Not Rescue 2, not Rescue, but *the* Rescue. It pisses off the guys from the other rescues, but the Rescue 2 guys love it. *Where do you work? The Rescue.* That's all. No need for explanation.

On the apparatus floor Ruvolo now gets an engraved identification tag over his cubby. Each new arrival pencils his name on a piece of masking tape atop his cubby, but when the official order comes through he gets to pull off the tape and put up a real sign. This row of coatracks and cubbyholes runs the length of the apparatus floor alongside the truck. The officers get the spots near the front, positioned to provide quick access to the cab. Midway down the row is Louis Valentino's cubby—boarded up forever by a memorial plaque that bears his name and helmet number. On the opposite, driver's side of the rig there are special lockers for the chauffeurs, one of their perks, in addition to the extra couple of bucks an hour they make.

The cubbies are where all the men keep their fire gear, which always smells like smoke and sweat. The guys take a certain pride in the lingering stink of their stuff—it means they've been busy. The only time they try to wash out the aroma of combustion is when they head home to their wives and kids. But, no matter how hard they try, the smoke smell never entirely disappears.

The helmets stowed in the cubbyholes record the experience of each fireman. Those of the young guys are new and still shiny, built from a lightweight, hardened plastic. Now they come in standard sizes. But in the old days there was a hatmaker who actually visited the fire academy and took elaborate measurements in order to produce helmets to form-fit each head. Many of these older leather helmets are still in use, though burned to a crisp and flaking apart from years of multiple alarms.

The senior men's battered helmets are their badges of honor, proud records of all the hell their owners have been through. Guys have been known to refuse to upgrade old helmets damaged beyond repair just because they want to hold on to the memories of every scratch, nick, and burn. The material for the helmet has changed in recent times, but its basic design remains the same as in the old days:

the wide brim of the helmet extends about eight inches in the rear. This rear extension helps hot water to flow away from the face and cascade harmlessly behind the fireman. On the front is a large white 2 on a blue background, blackened beyond recognition if it's a senior guy's helmet. There is also a plastic eye shield that folds down from the helmet, but no one ever uses it.

The cubbyholes also hold the personal items the men carry out of superstition, utility, or a combination of both. Lincoln wears a large hunting knife sheathed across his chest. He figures the knife might help him cut somebody out in a rescue, and it is always with him at a job. All rescue firefighters pack a small, tightly wound coil of rope, called a personal rope, used to bail out of burning buildings. The theory is that if a fireman is on an upper floor and that floor lights up, he can tie the rope off and slide down to safety. The men are well aware that it can take longer to rig the rope than it does for a fire to find them, just as they realize that a blaze can easily burn through the rope. But these facts don't seem to deter any of the guys from carrying their personal ropes. The ropes serve as a reminder that, no matter how bad things seem, there has to be a way out.

I N THE WEEKS THAT FOLLOW, as Bob and Ruvolo log hours together in the front of the rig, driving from fire to fire, pizzeria to pizzeria, in search of the perfect blaze or the perfect pie, they get to know each other. They come from the same little piece of Brooklyn, a string of old Italian neighborhoods in the southern part of the borough. When Bob mentions Cookie's, a bar he used to hang out at, the captain knows the place well. When Ruvolo picks Schultz's Deli in Flatbush for lunch, it turns out to be one of Bob's favorite places in the neighborhood. Working together one Sunday afternoon, they invent a route through Bergen Beach that they dub "the loop," which takes them on one easy circle to pick up homemade pasta, fresh baked bread, and great *sfogliatelle*.

The most remarkable thing is what's left unsaid, but understood, between them. Coming from the same place, getting on the job at the

TOM DOWNEY

same time, and having worked the same neighborhoods, they have a web of references that is like a private language. Just as some twins claim a deep unspoken communication, so do Bob and the captain. They trade laughs based on jokes that are never uttered out loud. Up front in the cab they laugh a lot: at themselves, at other firemen, at people on the street.

Like Bob, Ruvolo also moved his family out of Brooklyn to a more affordable neighborhood. But Ruvolo chose Staten Island, the odd man out of the four outer boroughs. While Brooklyn, Queens, and the Bronx have seen an influx of newly arriving immigrants, large swaths of Staten Island are havens for white working-class people fleeing Brooklyn. Sometimes it feels like whole Brooklyn neighborhoods have simply relocated ten miles west, across the river, in Staten Island.

Staten Island is a closer approximation of the old Brooklyn than Bob's Long Island will ever be. Where Bob's heritage dissolves in the melding of so many ethnicities, Ruvolo's neighborhood is proudly Italian-American. His home is near an Italian pork shop, a *pasticceria*, and a real Italian pizzeria, not one run by Albanians. Every other shop seems to be a *salumeria* or a tiny espresso place with plastic chairs out front, where old men sit watching the cars go by.

Ruvolo could pick up extra work as a carpenter or a plumber, or any one of the other trades he has mastered over the years. But he doesn't want to. His vocation is fire officer, and that's one thing he is clear about: even if the city won't pay enough to make it easy for him to live as a fire captain, and not as a part-time carpenter, plumber, or window washer, Ruvolo is determined to do it, to stretch his salary and make it work. After a few weeks of listening to his austere procla- mations about money, maintaining strict discipline over his young daughters, Jennifer and Victoria, and refusing to splurge on cable TV, the captain's family earns a new nickname at Rescue 2. He is the pio- neer member in a religious community that the men in the kitchen call "the Staten Island Amish."

Ruvolo is thrifty with himself, but he can't say no to his two girls; he pampers them. The only indulgences he permits himself are send-

62

ing out his light blue officer shirts to be washed and pressed every week and buying Nat Sherman cigars. He likes to unwrap a freshly laundered shirt at the start of each tour, even if it will be smelly and sweat-soaked after a couple of hours. He loves to savor a nice cigar in the kitchen after a big Sunday meal. Others opt for Cubans, but Ruvolo won't touch those commie cancer sticks.

By the first hint of summer, Bob has become the captain's most sympathetic listener. He agrees with most of Ruvolo's rants and raves about Brooklyn, politics, and morality. One day up in the cab he tells Ruvolo about his own strategy for dealing with his daughter's prospective boyfriends. At first, Ruvolo doesn't even want to listen; his older daughter is only fourteen, and he doesn't even want to think about boyfriends. But then he relents and lets Bob explain.

"I open the door nice and polite," says Bob. "I'm all smiles. If we're alone, I talk to him right then. If not, I ask him if I could have a word alone with him. Then I lay it out real simple. I look him right in the eye and tell the kid nice and slow, 'Anything you do to her, I will do to you.' It takes him a couple of seconds to process, but it works every time."

The growing bond between Ruvolo and Bob is the captain's first signal that he might actually make it at Rescue 2 after all. The captain's chauffeur acts as an informal liaison between the captain and his men. "You're pushing people a little too hard," Bob will tell Ruvolo, or "The men want you to know about this."

The two of them spend hours each day in the cab, sometimes during the frenzied rush to a job, other times lazily cruising for the meal. The cab is the only place to have a truly private conversation. The guys in the back can't hear what's being said, so Ruvolo and Bob can talk about what needs to get done: who needs a kick in the ass or some gentler encouragement. The only other place to have this kind of conversation is up in the office of the firehouse. But once a man goes up there to have a "private" talk, the other four guys on duty know about it. And they can't resist trying to find out what's being said. Once a tenacious eavesdropper set up a ladder in front of the building and climbed up to the captain's window to listen. More commonly, somebody will be hoisted up on top of the lockers and pry open the vent. But, in the front of the cab, doing his job, Bob is above suspicion. It's the one chance he has to talk privately with the captain and let him know what's happening with the men.

Now, when Bob hears someone bad-mouthing the new captain in the kitchen, he says, "This captain has done a hell of a lot for us in just a couple of months." And the rest of the men shut up. They are not ready to concede that Ruvolo might actually work out. But they don't dispute it—and these guys dispute almost everything.

FROM THE VERY BEGINNING of his career as a fireman, Ruvolo knew he wanted to be an officer. He was ambitious, but he also needed the raise badly. He started on the FDNY in 1978 at a salary of only $14,600 a year, and he and his wife, Janet, knew they wanted to have children. They also hoped that Janet could stay at home when the kids were young. Ruvolo posted a sign right above his study desk that read "A million dollars." He thought that if he studied hard and got promoted, he might make an extra million dollars over the course of a long career.

In the early morning hours before a tour, from 5 to 8 AM, Ruvolo would study for the lieutenant's test. Other firemen who had already aced the exam had organized their own Princeton Review–style cram course. Each week, Ruvolo trudged to one of these courses and then to a friend's house for a study group. At his firehouse the older firemen derisively branded Ruvolo and the other upwardly mobile young firemen "students." Some of these veterans had never wanted to become lieutenants; others hadn't been able to pass the test. Ruvolo knew he would. He had to.

After a few years of studying he also felt that he was reaching the point where he might actually make an effective officer. Early in his career, he had served at multiple fires each night and learned the job quickly. Fires were plentiful then. In one fifteen-hour night tour his company had thirty-two runs: five all-hands and a second alarm— a huge night of fire duty. Ruvolo came to know the buildings of Brooklyn, the behavior of fires, how to know when it was too hot, how to determine if a fire was getting dangerous. When the lieutenant's test came, Ruvolo passed with a high mark, an 89. But he had to wait four years to be promoted because he lacked seniority.

When he finally made lieutenant, leading five firemen didn't feel natural at all. He came out of a firehouse of veterans, where, even by the time he left to become a lieutenant, he still had less experience than many of the other firemen. When he went to talk to his captain about becoming an officer, the guy had told him, "Being an officer is like wearing a pair of new shoes. At first they hurt; then you get used to them and they feel good."

With his promotion to lieutenant in 1986, Ruvolo got a higher salary, his own office with a bed, and the chance to lead five men into a fire. For Ruvolo, the hardest thing about becoming an officer was the new responsibility to make split-second decisions that could lead to the death of a fireman under his command. He makes a constant and instantaneous calculation of risk versus benefit. But the terms of the calculation are never explicitly spelled out. When should Ruvolo risk his men's lives to save a victim? When should he restrain them even though somebody might be dying inside? The answers to these questions are intentionally left vague in department regulations, to be answered on scene with a combination of instinct and intelligence. And even after the fire is out, the civilian dead or alive, the fireman injured or unharmed, it might still be impossible to know if he made the right call.

At Ruvolo's first real fire as a lieutenant, he felt embarrassed by his authority and was reluctant to boss his men around. But then he looked at the fire raging in the basement below them and looked at the men he was going in with: inexperienced rookies with just a couple of years on the job. He had to lead them—otherwise they'd get eaten up in there. So he barked out the same commands his officers had given him, "Steady on the line. Open up the nozzle. Sweep the floor. Push in farther. Get in there. You got it, kid."

At first he heard a tremor in his own voice and wondered if they would listen to him. But they did as they were told, and Ruvolo realized that his authority comforted the men. They needed him to lead them. As he followed his nozzle man into the basement, still barking out orders, he panicked. *Where are my guys?* he thought. *Do I have everybody? Where's that probie?*

He shouted for the others. "Where are you, kid?" The probie was right next to him, thank God. "Stick by my side," Ruvolo said. Then he gave orders to the nozzle man: "Hit it higher, chase it back."

Now, ten years later, Ruvolo's hesitation has disappeared, but that little twinge of panic is never entirely gone. It's the price of being an officer. That's why he looks back at his days as a firefighter with such nostalgia. Those were the best years of his career and some of the best years of his life. Worrying not only about himself but about five other men is a life-changing responsibility. It might be more challenging to be an officer, but it's not nearly as much fun.

In the Police Department, promotion gives an officer the chance to retreat permanently behind a desk, but officers in the Fire Department are exposed to as much danger as their men. As captain of Rescue 2, Ruvolo sees much more action than captains of ordinary houses. Rescue 2 is charged with saving firemen. Ruvolo has to risk everything to pull a brother out of the flames. When a fireman signs on for this job in Brooklyn, he knows that if he's burning or suffocating the men of Rescue 2 will risk their lives to save him.

# THREE:

## THE SEVENTIES

**P**HIL RUVOLO LIVES WITH THE HISTORY of Rescue 2
every time he leads his company into a fire. Sometimes the house's
reputation brings good things, as when an old-school chief tells him,
"Do that rescue thing!" outside a brownstone ripping with fire. Or
when Chief Jack Pritchard, a Rescue 2 alum, says to his engine companies, "You got ten minutes to put it out, or I'm sending in Rescue 2."
But some guys in other houses cop an attitude about Rescue 2 based
solely on hearsay, without getting to know the men.

Firemen hold on to grudges forever; there are FDNY veterans
still walking around bad-mouthing Rescue 2 for things that happened
fifteen or twenty years ago. Memories are passed down from one generation to the next: fathers to sons, uncles to nephews, and senior men
to junior men. Every company has its legendary exploits: rescues,
close shaves, and wild rides. Everything a company has ever done
eventually becomes known to all. The identities of individual houses
have developed over the years, and even men with strong personalities often change, as time passes, to conform to the houses they
belong to.

When Phil Ruvolo walks into a fire, he's accountable not just for
his own actions, accomplishments, and mistakes, but for everything
his company has ever done: all the good and all the bad and everything in between. There's never any doubt about Rescue 2's firefighting

ability or courage, but a lot of guys in lesser-known houses are sick of big Rescue 2 egos. A common complaint is that Rescue 2 doesn't show the proper deference to chiefs or to men from other companies. This reputation really began in the early seventies or, to be more specific, in 1973, when then-captain Fred Gallagher began the transformation of his firehouse from a bunch of specialists into the equivalent of the FDNY's special forces; it was like going from being a good swimmer to a navy SEAL or, in Rescue 2 lingo, from being a proctologist to a real asshole.

FRED GALLAGHER had never wanted to join the FDNY. Growing up in the fifties in Flatbush, Brooklyn, he'd seen fire trucks roar by and had observed the neighborhood firemen's nightly procession from the firehouse to the local watering hole. Occasionally, if the flames were high, he had crowded onto the sidewalk with everyone else to gaze up at fires. But the fireman's job had never captured his imagination. Fred was a jock and thrived on competition: baseball, football, neighborhood brawls.

In the working-class sections of New York in the fifties almost every block had a vacant lot that served as the ball field for the neighborhood team. Fred's baseball career started in a dusty lot on his corner. As his team advanced in its league, Fred and his friends had the chance to travel farther and farther away from the neighborhood for games. At playoffs in the parade grounds, in Brooklyn's Prospect Park—a grand space with thirteen ball fields—he watched a great young hitter, Joe Torre, future manager of the Yankees. Later Fred played ball in a pregame show at Ebbets Field, just before the Dodgers game. He was a football standout too, and at seventeen joined a semipro football team run by a Runyonesque character named Pudgy Walsh.

Fortunately sports were virtually free, which was a good thing because growing up Fred didn't have much money for luxuries. He could afford the subway only if he was going to work or to school. The rest of the time he rode his bike. He almost never left the borough. An exotic outing was a family car trip in a noisy old clunker, to Valley

Stream State Park, on the edge of Queens. Sometimes he saved up sixty cents for a seat in the bleachers and pedaled over to Ebbet's Field to watch a Dodgers game. If Fred got there early, he could watch the players pile out of their beat-up jalopies; back then there was plenty of fame but no endorsement deals or inflated salaries for the workaday players.

After Fred graduated from high school, he got married and looked for a job. The neighborhood men worked hard to make ends meet: they were tradesmen with union jobs or civil servants always talking about the pensions they would get after giving the city twenty years of sweat and toil. But Fred wasn't having it. To him, those civil service jobs were boring, and the union racket was even worse. The civil service jobs were based on whom you knew and how you did on some kind of test, not what you did as a man. He landed a job as a bill collector in Harlem and ignored the announcements in the post office that the Fire Department exam was coming up. His mother-in-law, on the other hand, saw things differently. A civil service job for Fred meant a steady income forever for her daughter's family.

Fred was a young man with strong opinions and a quick temper, not the kind of guy who normally took advice from his mother-in-law. But, tough as he was, he was no match for this willful lady, who chased him down the night before the Fire Department test and told him, in no uncertain terms, that he *was* going to take that test the next day— or else. Fred had gone to Brooklyn Tech, the borough's magnet school, and he'd always aced his exams. The Fire Department's written test was a breeze, and the physical exam was even easier for a guy who was benching close to four hundred pounds and practicing baseball or football nearly every day of the year. Soon after he took the test, Fred was reporting to a local engine company for his first day of work.

It was 1960. He knew so little about the job that he showed up wearing a gray T-shirt—which was officer's attire in the days before there was a uniform. Surprised at the deference the older men showed him, Fred happily sat at the kitchen table all day, sipping coffee while the company labored around him. He didn't realize they thought he was an officer. Then someone said, "Where's the probie?"

"That's me," said Fred.

"Well, what the hell have you been doing sitting around all day long?" demanded one of the salty veterans.

That was the end of sitting around. Galllagher didn't mind because, after his first taste of firefighting, he was hooked. As an athlete he had always found a way to compete and excel; now he made everything into a matchup and tried to win. It didn't take long for him to decide that knocking down fires sure beat collecting bad debts in Harlem.

The only problem was that his company, Engine 210 in Fort Greene, Brooklyn, was slow. Bored stiff and ready for some action, Fred wanted to go somewhere busy. But he had to go through the tedious process of applying for an official transfer; he didn't have an in to help grease the wheels. When he put in for a truck company, for which the height requirement was five feet, ten inches, he got rejected; his official height was five feet, nine and three-quarter inches. Fred wasn't giving up that easily. He stormed down to the borough command after work and forced an old-timer sweeping the floors to measure him. Miraculously, the guy found that extra quarter of an inch somewhere and put through the transfer. Fred had a way of convincing people.

He secured a spot at Squad 3 in Bedford-Stuyvesant in 1962, a time when the city was just starting to see some real fires. At that time the city's three squads worked as additional manpower units; they gave relief to overworked units in busy areas. In the early sixties, most blazes were concentrated in a small group of neighborhoods that were going bad fast. Older residents were moving away every day, and newcomers had few roots or bonds in these communities. Brooklyn's Brownsville had been a working-class Jewish haven until the fifties, when vast tracts of public housing brought a new African-American population to its streets. These people who lived in enormous, impersonal new projects had little connection to where they lived. Poor and struggling, they had come to Brownsville out of desperation, and a few people acted desperately. Vacant buildings burned to the ground. Arsonists had a field day.

When places like Brownsville and East New York burned, urban decay spread to neighboring areas, and the fires followed. Bushwick started to burn, then Bed-Stuy and Crown Heights. Even Flatbush,

Fred's old stomping grounds, would follow a few years later. Fred found himself at the heart of the action—the place he had always intended to be. On the national news, the South Bronx was held up as the most prominent symbol of urban blight. But, in fact, the Bronx was so tiny that eventually nearly every vacant building there had burned to the ground. Brooklyn, many times larger, raged for years.

By the late sixties and early seventies, New York's ghettoes blazed with a kind of fury that had never been seen before. Crime, drugs, racial and ethnic tensions all meant more big fires. Most low-income neighborhoods had plenty of vacant buildings just waiting to light up. When a junkie went postal or a warming fire in winter got out of hand, these vacants burned. Back then every thug didn't pack a handgun, so a fire was also a good way to get revenge. Criminal and domestic rivalries could easily erupt into wars, with people burning down houses, torching apartments, and committing murder by arson. Arson was seldom prosecuted, and insurance money was so easy to collect that many failing businesses reached for the kerosene can rather than file for bankruptcy.

No Brooklyn fireman could ever forget the first time he arrived at a project fire and saw a whole family on the street, in front of their burning apartment, with trunks, television sets, and wardrobes packed to go. If a family's apartment in the projects burned down, that family jumped to the top of the list for a new, often better city apartment and got a thousand dollars for being burned out of their place. Some poor people were desperate enough to torch their own homes.

Arson was on the rise all over the nation, but the thing that set New York City apart was that the FDNY had to aggressively fight every single fire. In Florida or California, firemen could give up a fully involved house and let it burn if they knew no one was in there. That was the right call in areas where buildings stood far apart. But, because of New York's dense construction and uncertain squatter population, the FDNY had to fight every fire—or risk losing whole blocks and whole families to a blaze. A Brooklyn Fire Department that in 1960 had responded to fewer than 100,000 calls was responding to over 170,000 by 1978.

UNTIL THE SIXTIES, firefighting wasn't so different from other civil service jobs. The risks were greater and there was a special esprit de corps forged in fires, but for most firemen the department was simply a good way to make a living. But in the sixties a new myth was born in the FDNY: magic, mysterious, all-powerful *ghetto firemen* appeared. Unlike your usual humble civil servants, who punched time cards for twenty years before settling into staid retirements, the ghetto firemen were soldiers who risked their hides every day to make the city a safer place to live in. And they liked doing it. Ghetto firemen wanted to be where the action was, in the toughest, most anarchic neighborhoods of the city. The ghetto fireman—the white working-man's war hero—was born. But at the same time, as tensions rose in the city, some people in the ghettoes saw firemen, like cops, as *the man* and associated the FDNY with the repressive power structure of the rich and entitled.

Most ordinary people still admired firemen for risking their lives to protect the city's toughest neighborhoods. But some angry arsonists got personal. They didn't just set fires; they rigged booby traps to hurt firemen. They might cover a gaping hole that dropped into the basement with a piece of linoleum. Crawling in for a job, a fireman would feel the linoleum, think it was safe to enter, and then fall through. Other times criminals filled balloons with gasoline and left them in a fire building. When things got hot, the balloons would blow and catch firemen unawares. Suddenly, a fire that a guy thought was in front of him was chasing him from behind.

There was anger flowing across the streets of Brownsville and East New York, rage igniting vacant buildings in Bed-Stuy and pouring out of gasoline cans in Crown Heights. This was a war with real casualties, and the foot soldiers sent to fight the battles were almost all working-class white boys—firemen and cops who had their own class gripes and prejudices. After all, Madison Avenue families didn't send their sons to fight America's war in Saigon—or the one in Brownsville.

Work in the good neighborhoods didn't change much. In more upscale buildings, firemen still got treated as part of the municipal servant class. Before they went into swanky apartments to put out fires, owners often asked them to take off their dirty boots. A restaurant might ask a company to come back at 2 AM, after closing time, to clean out a clogged sewer line. When they were treated like that, the firemen simmered with resentment at the nerve of these rich pricks. Meanwhile, in the ghetto, busy companies might start off their night with a burning vacant down the block, then go to a shooting, an overdose, and a couple of torch jobs, before finishing off the evening with a second fire in the same building where the night had begun, fifteen hours before. The job wasn't your average civil service post anymore. It was more like your average jungle reconnaissance mission in 'Nam. When Vietnam vets stepped into a ghetto firehouse, they encountered the same macho intensity they had felt in the war.

To firemen like Fred Gallagher, the constant parade of human tragedy he witnessed every day in the ghettoes of Brooklyn created a desire to do the job better, faster, and more efficiently. God knows, people in the ghetto needed any help he could give them. When he saw hungry kids in the morning, waiting for the free meal at school, or when he stepped over a junkie to enter someone's home, he realized that people in Brownsville or Bed-Stuy really needed him—and, whether they liked it or not, it was white guys like Fred who were coming to help them. There were almost no firemen from Brownsville or East New York. And there were few black or Puerto Rican firemen at all. The department and its few black and Hispanic firefighters had been trying for years to get kids from these non-white neighborhoods to join. Most white firemen thought that the reason these kids wanted no part of the job was because they had seen it for real. The kids in Brownsville or East New York often watched exhausted firemen come out steaming, puke in the street, and collapse on their sidewalk. But for a white kid from a better part of the city, who didn't know any better and had never spent much time in the ghetto, the job held an exotic appeal. It was a passport to a place he would never otherwise go.

New York firemen traditionally were from a different world than the people they served. They had often grown up in the same handful of Irish and Italian neighborhoods that dotted the outer boroughs. They knew each other from Catholic Youth Organization basketball games, from parochial school, and from their neighborhood bars. The FDNY was not just a job; it was the official caste profession of working-class Irish and Italian Americans—an outgrowth of Catholicism, patriotism, and civic pride. It promised a steady salary back in the days when that was hard to come by for a working stiff from the outer boroughs. The job earned you the deep respect of older guys in your neighborhood, and it meant lifelong camaraderie: camping trips, dinners, and nights at the local watering hole with a group of men you knew and loved.

So it was mostly Irish and Italian guys who wound up fighting fires in the Brooklyn ghettoes, and many of them loved it. Their generation was too young for World War II or Korea; this was their glorious war. And it wasn't just the firemen who were addicted to the job. Even the men at the fire dispatcher's office, the guys manning the radios, were hard-core buffs. They collected stats about companies and captains as if they were collecting numbers on ball teams and pitchers. During the sixties and seventies, before computers streamlined the process, alarms were still transmitted as pneumatic signals sent through an old-fashioned mechanical system. When a job came up, the veteran dispatchers, who had memorized every detail about every alarm box in Brooklyn, knew exactly which companies to send without even consulting the chart. They rattled off numbers like idiot savants: *"Box 1994—first-due assignment is Engine 255 and Ladder 147. If it goes to a 10-75, they get 310, 309, 157, and the 41 chief."*

To the uninitiated listener, the numbers were gibberish; to the dispatchers, each number let them imagine a group of guys they knew suiting up and heading out for battle. Some dispatchers, like Warren Fuchs who buffed at Rescue 2 on his days off, were so good that they could tell exactly where a company was in Brooklyn and alert them when another rig was coming their way. Warren might get on the radio and say, *"Brooklyn to Rescue 2, watch it when you turn onto Troy because you've got 123 coming your way."*

Chaos reigned in the sixties and seventies. False alarms were rampant. Certain boxes were guaranteed to trigger at least five or six false alarms a day. In some cases it was absolutely predictable: a box outside a school was pulled every day after classes let out, except on holidays when the kids weren't there. Other boxes usually meant real fire duty. A box pulled in the quiet streets of Bergen Beach almost always meant a fire, not a false alarm.

For some firemen's wives and mothers-in-law, the increase in fires didn't seem so terrific: a job that had been only remotely risky now came with a near certainty of injury. Every guy in one of these busy companies eventually got a concussion, a broken bone, or a nasty burn. But for Fred Gallagher and men like him, the war years brought the challenges and thrills of a lifetime. Now the professional ghetto fireman was playing in the World Series of fires every night. You had a chance to prove yourself every minute, and you always had a good story to tell when the next shift came to work.

AFTER A COUPLE OF YEARS at Squad 3 in Bed-Stuy, Gallagher joined Rescue 2 as a fireman in 1963. The hotshot officer there was Dick Hamilton, a brave lieutenant whose exploits had been detailed in a Playboy Books biography. In his first year at Rescue 2, Gallagher watched as Hamilton climbed under a horrific train wreck and administered morphine shots to an injured conductor, while the other Rescue 2 men worked to untangle the victim from a mass of metal. That made a big impression on a young fireman. Hamilton had no fear.

Gallagher took the lieutenant's test in 1967, then the captain's test in 1968. He scored second in the city on the lieutenant's test and first on the captain's. When he was promoted to lieutenant, at age thirty, he went to Harlem, which at that time was as busy as anywhere in Brooklyn or the Bronx. It was all the action a young ghetto fire officer could want: the city continued to burn around him.

One night, at a fire in a pool hall, he faced his first test as an officer. The pool hall was sealed in the back, making it impossible to vent. So the fire had only one way to blast out: the front entrance, where

Gallagher's engine company was pushing in. He and his men were plastered to the ground, with fire rolling over their heads.

Gallagher had to make the call: bail out, or push in and hope that the flames didn't make it down to them. The guys looked to him for an order. He had just a few seconds to decide. He put it all together in his mind: only fifteen or twenty feet stood between them and the seat of the fire; it was hot, but his ears weren't burning yet. They could make it. He told his men to get in there. They dug in and advanced, inch by inch. As they got in farther, the flames banked down closer to them. But they made a last push around the corner to knock them down, and they still had over a foot between their heads and the fire, more than enough space to feel safe. Putting that out was the beginning of the Gallagher legend.

As he commanded his men at more fires, decisions like that one got much easier to make. Over the years, he became known as an officer who hardly ever told his men to retreat. Years later he could look back on his career and say that he'd given up only one row house.

When he made captain in 1969, Gallagher was sent right into the firestorm. He got assigned to an engine company in a "double house," one that includes both an engine and a truck company, in the Brownsville/East New York section of Brooklyn. After captaining the engine for a while, Gallagher crossed the floor of the double house and took over as captain of the truck. Ladder 103, his new command, was a colorful place. Long, unkempt beards and mustaches were popular, and there was one guy with so much facial hair he was called "the Animal." Some of the older guys at 103 had served in World War II. One day a new lieutenant strolled into 103 and recognized a chief from their days as fellow Allied prisoners in a German POW camp.

It was at 103 that Gallagher met Tom Dillon, another kid from St. Thomas Aquinas in Flatbush, Gallagher's childhood parish. Tom quickly became his regular chauffeur and best friend; it was a partnership that would last a decade.

Tom had left Flatbush at age fifteen, after he was orphaned. He had gone to Queens to live with his aunt and uncle. He quit school at sixteen, briefly worked as a wood butcher (a large-scale carpenter), then joined the marines. He had to fend for himself from an early age, and that made him tough. The pain and suffering of the job hardly penetrated Tom's thick skin—until the night when he carried a dead little girl out of a fire dressed in a pink nightgown that looked just like his daughter's.

Officially, Ladder 103 did about eight thousand runs a year, but that was back in the day when runs were recorded on little scribbled scraps of paper and stuffed into an officer's pocket. It was a miracle that any scrap lasted a whole tour without getting lost or burned to a crisp. Some slips deliberately went missing. If a company worked more than twenty runs before midnight, they had to go to a slow district of the city the following night to rest. Gallagher made sure that this never happened to his company. When midnight came around, he always had fewer than twenty scraps of paper left in his pocket.

All Brooklyn engine companies had three-digit numbers, but there were nights when the borough was so busy that Gallagher never saw a three-digit engine at fires. The dispatchers would frantically relocate companies from the slower parts of Manhattan and Queens to protect Brooklyn. Gallagher saw all kinds of triumph and tragedy, but one of the most memorable runs at 103 was right on his street. Gallagher and his men were heading back from a job when the dispatcher ordered them to another box. As they raced back, Gallagher listened to the address and realized it was his firehouse. By the time they arrived, flames were shooting out of the second-floor window, right next to the kitchen. Somebody remembered he had run out the door with supper on the stove. The chauffeur turned to Gallagher with a grin and said, "I think we got a job."

"I hope we still do," Gallagher quipped.

After a tough push into their own firehouse and up the stairs to the kitchen, they put it out. But the second floor was in ruins. The roof was open to the sky, and there were charred remains all over the place. The men kept meaning to at least fix the ceiling, but every time they tried, another run came in, and for months every time Gallagher lay down on his cot he could see the moon through the hole in his ceiling.

The chief's test was coming up in the early seventies and the

salary jump from captain to chief was the biggest increase on the pro-
motional ladder. With four kids born in four years, Gallagher needed
the money, so he put in his papers for the chief's test. But Chief of
Department John T. O'Hagan noticed that Gallagher and another
young whiz-kid captain were poised to make chief with hardly any
years on the job as captain. O'Hagan rescheduled the exam, effec-
tively making them ineligible until next time around, five or so years
down the road. O'Hagan had other plans for Fred Gallagher; Gallagher would captain a rescue company next, and he could become a
chief after he mastered that.

THE NATION'S FIRST RESCUE COMPANY, New York's Res-
cue 1, was founded in 1915 after a spate of fires made it obvious that
the FDNY needed more modern, high-tech equipment. In 1912, at a
fire in the Equitable Building, firemen spent hours working with a
hacksaw to free people trapped behind metal bars. At another fire in
the same period, poisonous gases had repeatedly knocked firemen
unconscious. The department sent a chief to Germany and France to
see what equipment was available for the newly formed Rescue Company 1. One of his first acquisitions was a torch that could melt metal
bars in minutes.

In the early 1900s, the FDNY's basic strategy was to throw as
many men as possible at a major fire and accept that a good percentage of them would pass out and have to be dragged to safety by the
men behind them. Most of the time this strategy worked fine, but in
fires where there were toxic gases like ammonia or carbon monoxide
in the air, firemen could easily die. The Draeger smoke helmet from
Germany, one of the first items carried on the Rescue 1 truck, allowed
men to breathe inside a fire for a few precious minutes. This helmet
and the other special equipment the company used were crude pro-
totypes of the sophisticated technology that would blossom later in
the century, but at many fires in the early days, the smoke helmet or
cutting torch did the job.

From its founding in 1925 until 1973, Rescue 2 went strictly by the book, responding when they were expressly called to a fire or emergency. Otherwise, like every other FDNY company, they sat in the firehouse and heard fires come in on the radio. They didn't leave quarters until they got a proper invite. Rescue 2 developed a reputation as a boutique firehouse that only went to huge fires or strange jobs—not what it would eventually become: a workhorse company called to every fire in the borough. Before 1973, even when Rescue 2 did go to fires, the chief on scene would often just order the company to stand fast.

Rescue 2 had some strong officers and firemen, and won countless accolades. But there were definitely some duds in the company. One notorious lieutenant, a beneficiary of old-fashioned New York nepotism, couldn't even remember the names of the men whose lives he risked each tour. He called everyone Joe or Bob. This lieutenant read out loud an official department edict against firehouse alcohol consumption with an open beer can in one hand and the order in the other. Later, when Fred Gallagher went after the guy, the chief of department told him the officer "was the best axman [he] ever saw."

"Well," said Gallagher, "he's not swinging an ax anymore."

Rescue 2 was supposed to be the best firehouse in Brooklyn, but as it was, most ghetto firemen wanted to be somewhere else. The place needed to change, and so in 1973 the chief of department, in charge of running the whole FDNY, chose Gallagher as the man to transform Rescue 2. One night the Brooklyn borough commander came to see Gallagher in his office. "You're going to Rescue 2," the commander told him. "We've got a problem over there."

When Fred Gallagher returned to Rescue 2 in 1973, as its captain, he came to lead a company that was supposed to be the special forces of the department—the guys who saved other firemen. But that mission was endangered by the fact that they weren't working much at fires. Who would rely on these men to perform under extraordinary circumstances if they weren't out there fighting fires every day, performing under ordinary ones?

Gallagher wasn't a guy who could bear to just wait for something to happen. He was impatient, driven as an adult by the same irrepressible energy that, as a child, had catapulted him around the bases or across the goal line. Not long after his arrival, when Rescue 2 was called to a Brooklyn jail for an escaped inmate trapped on a ledge, Gallagher showed just how impatient he was. The police were standing around strategizing. Meanwhile Gallagher climbed up next to the escapee and said: "Listen, this is not the Police Department, it's the Fire Department. If you do anything funny, I'm gonna throw your ass right off this building." They put the inmate in a stretcher and took him down without a fight.

Gallagher couldn't stand being bored. He had itched for action at his first firehouse—and later found it in East New York. Now, at Rescue 2, he was again eager for excitement. For a month or two, he listened to the chiefs on scene who told him to keep his company out of the fire or simply go back to quarters. Underneath his blood boiled. What was the point of commanding an elite unit in the busiest fire locale in the world if they didn't get to fight fires?

Two months into the assignment, he couldn't stand the boredom and frustration any longer. *I'm the captain of this company,* he thought to himself, *and I can make it into whatever I want.* He called up his old buddy and driver, Tom Dillon, and got him into Rescue 2. He brought over some of the best firemen from East New York. And then Gallagher and his new Rescue 2 started responding to every working fire in Brooklyn. The old Rescue 2 had never done that, but there was no arguing with Gallagher. Talking back to the new captain could get you pinned in a headlock or taken down to the basement for a one-on-one screaming session. Fred Gallagher now knew his mission and no one was going to stop him. His Rescue 2 was going to be at every fire, big or small, and Gallagher's men were going to work those fires, not stand around.

Gallagher blew into Rescue 2 like a tornado, and before long, many of the guys from the old Rescue 2 were putting in their transfer or retirement papers. Some guys the new captain was sad to see go; others he booted out the door. Within two years he had filled the ranks with men he handpicked from busy Brooklyn ghetto companies. He had a couple of requirements: The men had to be under forty years old. (It stretched to forty-one and forty-two when Fred himself got older.) They should have a trade, like welding or carpentry, that they could use at emergencies. Most important, though, he wanted people like him: guys who lived to fight fires.

He circulated a document to all the battalion chiefs in Brooklyn that described the mission of Rescue 2. Rescue 2's job was to provide six outstanding firefighters at every working fire in the borough, men who would make sure that these fires were put out swiftly. Rescue 2 would protect other firemen not by standing outside and waiting for a catastrophe, but by being in there, hard at work, ready to pull a brother out of the flames.

In the ghetto areas, working shoulder to shoulder with experienced firemen who had egos the equal of Gallagher's own, Rescue 2 hardly ever got the hose line. Ghetto companies didn't have many probies. These companies were typically staffed by senior men who knew what to do. But now and then, an inexperienced Brooklyn company with new officers and green firefighters went up against a tough job, and there was trouble. The longer a fire burned the more lives lost or put in danger, and the more buildings forfeited to flames. Sometimes all it took was a little coaching and encouragement by Rescue 2 to get an inexperienced engine company to move the line where it needed to be. Sometimes an engine company had already taken a shellacking, long before Rescue 2 arrived, and needed relief. Other times Rescue 2 would arrive and find the line abandoned; then they'd pick it up and put out the fire. Gallagher had six ghetto-hardened firemen on duty twenty-four hours a day, seven days a week, to take over for a struggling engine when they needed to.

In the Fire Department, progress is measured in inches: A hose line moving forward, even slowly, is getting closer to the source of the flames. "Making a room" means pushing the line into a room engulfed in flames and knocking them down. Often the chief out on the street will read the engine company's progress by resting his foot on the hose line and tracking how fast it moves.

But Gallagher went right inside the fire building and hovered over the engine company's shoulders. He and another Rescue 2 guy would get behind the nozzle man and hump hose. If the engine moved in fast and killed the fire, the rescue guys wouldn't say a thing, except maybe "good job," after it was all over. But if there was hesitation, or if the engine was a little out of practice, Gallagher was there as coach and drill instructor. Most of the time his presence alone was enough to make the line move faster. When this screaming giant was on a firefighter's back, there was nowhere to hide. It was drag the nozzle forward or get run over by the train. One way or another, Gallagher figured, that fire was going out.

In smoky hallways a fireman could hardly see his own hand, much less determine whether the guy next to him was from Rescue 2 or from Mars. If someone nudged up behind him and offered to take the line when he was getting pounded, so what? But then, after the fire, standing on the sidewalk catching his breath, he'd think to himself: *Wait a second, who the hell took the line?* Then it dawned on him. *Fuck. It was the Rescue.* The other companies hated the new Rescue 2. Gallagher loved what he had created.

He transformed Rescue 2 into the most active firefighting unit in the city. His strategy was simple and direct. Fighting a fire is loosely organized chaos, with engine men battling the blaze, truckies venting and looking for victims, and chiefs in charge on the street, coordinating the attack. Gallagher focused on one thing: knocking down the fire. When a fireman pulls up to a burning building and sees victims hanging out windows and parents screaming for their children, his instinct is to go for the grabs. But, Gallagher reasoned, if the engine knocks down the fire right away, those rescues are unnecessary. Everyone is safe. He stole his motto from Visine: "If you get the red out, you see better." If the chief told Rescue 2 to go somewhere other than behind the nozzle, Gallagher would dispatch only his right-hand man, Tom Dillon, on the chief's errand and trust Tom to take care of the whole assignment. Then he'd focus the other four guys on the fire. Gallagher would always get intimate with the engine company, crawling right up behind them. It wasn't uncommon to get a stray boot in the face if he got too close, but Gallagher didn't care; he was right where he wanted to be.

After a fire, if they weren't rushing across the borough to another alarm, the men of the new Rescue 2 would gather round Gallagher's desk in his upstairs office to talk things over. He usually started by trying to gently persuade his men to do things *his* way the next time they fought a fire. If they didn't agree he moved on to less gentle forms of persuasion. After some preliminary discussion Tom Dillon would tell Gallagher, "I think I understand. I get it."

"You do?" asked Gallagher.

"Yeah. We'll compromise . . . we'll do it your way." Everyone would burst out laughing.

Tom was a soft-spoken and mild-mannered counterpoint to Gallagher. From the age of sixteen on he had worked like a dog, always juggling two or three jobs and almost never having a day off. Often he would come to work so knocked out from his previous firehouse shift and his side job delivering oil that he would fall asleep while driving the rig. "You dumb Irishman," Gallagher would yell at him. That always woke him up. Sometimes Warren Fuchs would be crouched down in the cab, riding with them. When Tom got lost, he would ask Warren for directions. Warren would peek out the window from down on the floor and direct Tom to the fire.

On February 27, 1975, Tom was posted outside a fire room at the legendary New York Telephone Fire, the biggest and longest burning fire the city had ever seen. He was there as backup in case the firefighters needed help inside. It was so smoky that Tom had to wear a mask or suffocate on toxic fumes. He didn't have anything to do but wait, so he fell asleep in a chair, dozing until the other men came out. Fortunately, the air tank had a loud alarm that would wake him up when his air got low. In the midst of his nap, Tom awoke to some swift kicks from another Rescue 2 firefighter. Tom never heard the end of that one. The guys couldn't believe he dozed off in the middle of the biggest fire in New York history.

Years later the men who worked the phone company fire would wonder about all the stuff they inhaled there. Even wearing their masks, each man still got a heavy dose of toxins. An unusually large number of the men who worked the Phone Company Fire developed cancer later in life. But there was no way to tell if it was that fire in particular that made them sick; all those guys had been breathing carcinogens for decades.

Most of Gallagher's recruits were guys he knew from captaining Ladder 103 or guys who got strong referrals from other Rescue 2 firemen. But two of the men who became his all-stars got into the company through the back door, using their weight to get a spot and circumventing the captain. One might have expected these guys to be subpar, but, in fact, Jack Pritchard and Pete Bondy went on to become legends of Brooklyn firefighting.

Jack Pritchard was one of the most naturally gifted firemen of his generation, unstoppable at a blaze. At the firehouse, though, he was an officer's worst nightmare, a nonstop joker and roughhouser. A lithe five-foot-five, graced with a maniacal grin, Jack was dwarfed by Gallagher, who still looked like a Brooklyn lineman. Jack loved to start trouble and stir things up whenever he could. He was a terror in the firehouse, provoking and insulting everyone, always itching for a brawl. Jack was wound up like a tight spring. Sometimes Gallagher would intervene on his behalf when he got into fights, but that made Jack even angrier. He could fight his own battles, thank you. Big guys who messed with him found themselves up against the toughest, strongest 150-pounder in Brooklyn. But whenever he plunged into a fire, he stopped messing around and he became all business.

Before he maneuvered his way into the company, Jack wanted Rescue 2 bad, but Gallagher told him there were about fifty more qualified guys ahead of him. Then Jack got a break: the unit he was working in, Squad 4, had to cut down its roster of men. Jack knew that Squad 4's captain had pull with the top brass because the captain had chauffeured the Brooklyn borough commander earlier in his career. So Jack approached Squad 4's captain and made a proposal: "I'll leave here voluntarily if you can get me a spot at Rescue 2."

When Jack reported to Rescue 2, he walked straight up to the captain's office. Gallagher stared down Pritchard as he walked through the door and told him point-blank: "I didn't ask for you. I don't want you. And I don't know how you got here." Jack scowled back at him. They pissed each other off immediately.

But there was something in the captain that made Jack, who never backed down from or made peace with anyone, want to reconcile with

Gallagher. Jack walked back up to the captain's office a few minutes later and said, "Look, we started things off all wrong. Gimme six months to prove myself, and if you don't think I'm up to it after that, I'll leave voluntarily." Gallagher needed only one fire with Jack to know he was keeping him.

From that fire on, Fred never had to worry about Jack's abilities inside a fire. The only thing that concerned him was his suicidal tendencies when he walked into a blaze. If there was a fireman or civilian in trouble, there was no holding Jack back. In 1975, a year after he arrived at the company, Jack made one of his first daring grabs at Rescue 2. Many more would follow in his seven years at the company. That night the fire burned in a three-story building with a storefront church on the ground floor. People on the street told the firemen that a retarded boy, who lived on the second floor, was still trapped in the building. The first and second floors were consumed by fire and the stairs were impassable. The skylight at the top of the stairs had been shattered, venting the fire. That pulled the smoke out of the building and made it easier for firemen inside to search and survive, but it also turned the stairs into a chimney and sent smoke and flames leaping up the stairs to the roof.

Neither the stairs nor the roof was an option, so Jack climbed a ladder to the third floor and went in a window. To fit through the window he had to take off his air tank, so now he faced the fire without a mask. Thinking that the boy might be trapped in the hallway, Jack made a quick pass through the apartment, then crawled into the hall. By the time he hit the hallway, it was pulsing with smoke and fire. The wooden banisters were burning. Jack went for the inside ladder to the roof. Maybe the boy was trying to get out that way. On the crawl to the ladder he ran into the boy. Reaching for him, Jack immediately felt the heat. The kid was on fire. Grabbing the boy, Jack rapidly patted out the flames with his gloved hands and sprinted for the apartment from which he'd entered the hallway. But the door had locked behind him. He threw all his weight against the door. It didn't budge. If he stayed out in the burning hallway for another minute, he and the kid were both going to burn to death. There was nothing left but to wrap the kid in his arms and dive over the banister, down to the second floor.

He landed with a crash on the stairs. The second floor was just as bad. Again he took hold of the boy and dove over the banister to the ground floor, where a stunned nozzle man turned his hose on the boy and on Jack. They were carried out the front door and laid out on the sidewalk. Firemen wet down their burned bodies while they waited for an ambulance to arrive.

The kid lived for only a day. Jack always regretted that he wasn't able to save him, but he also felt that his rescue meant one more day for the boy to make peace with his family and with the world.

Back in the kitchen Jack found that the guys were as brutal as always. "You killed him, Jack," they said. "What do you expect when you throw a kid down the stairs?" Jack shrugged it off. He knew it was just their way of dealing with heartbreak and hassles. Other people might cry, pray, or plunge into a deep depression. The Rescue 2 guys busted balls. Jack spent a few weeks in the burn center. The minute he got out, he was itching to go to another fire. He got some hardware for that one, the Brummer Medal, which came with an $850 prize. But even then, with a year at Rescue 2 and a medal pinned to his dress uniform, Jack was dreaming of winning the Gordon Bennett, the highest honor the department could bestow. Jobs like that earned Jack Pritchard a reputation in Brooklyn—now everyone knew he wasn't just a wise guy. He was one tough fireman with a gift for making grabs when they mattered.

PETE BONDY was a quiet kid who had come to Rescue 2 from a slow ladder company in Greenwich Village. It was a fluke that he ever made it to the company. Pete had been trying to transfer to Rescue 2— or anywhere—for years, but that had been impossible until one night a guy called up from the commissioner's office. He was looking to move somebody into Ladder 5. Pete was sitting next to the officer who took the call and when he heard the exchange, Pete volunteered to give up his slot in Ladder 5. The commissioner's guy got his transfer in and Pete got his transfer out to Rescue 2.

When they went into their first big job together, Gallagher nudged Pete to take the hose. That was the last time he ever had to tell Pete anything but "hold back" inside a fire. Pete let out a yelp of joy and chased the fire down the hallway. When Gallagher saw just how aggressive Pete was in fires, he thought to himself, *I've created a monster.*

Once, at a fire in a vacant building, Pete paused, waiting to open the hose line and put out the fire. "It's my birthday, Jack," he said. To celebrate, he wanted the fun of knocking down a bigger fire. He wanted it to get a little warmer before he opened up the hose.

But when it started to get really hot, Jack yelled from behind, "Enough with your fucking birthday candle. Put it out."

In a fire, Jack and Pete were known to get competitive. One was more fearless than the next, and in the end they both got burned, cut, or injured a lot, though neither ever went sick if he could possibly avoid it. Jack and Pete wore their burns like medals, and their upper bodies were covered with scar tissue. The skin was like something you'd see in a medical textbook. Neither could ever be kept in the hospital long enough to get properly healed.

Pete would treat almost anything with a couple of aspirin or a little burn cream. After getting cooked at a job, he would come back to the firehouse, clip off the burned, dead skin with a scissor, and spread a little Silvadene burn ointment on the wound. Then he'd be ready for the next one.

Once Pete went to the floor above a fire and got caught up there in a flash over. He jumped out of a second-story window, on fire, and was captured in that pose by a *Daily News* photographer. Pete looked like a fireball shooting out of the flames. After that one, Gallagher told him to get his ass over to the hospital. Pete considered it but decided instead to take two aspirin and see how he felt in the morning. He was back the next day, limping into work, carrying hot bagels in his one good hand.

Fred Gallagher was an old-school father figure who could be stern if you fucked up, but he was also a friend to his men. The guys loved him. He would join them in the kitchen for food fights and fist-fights, then lead them into a fire and demand that they risk their lives. He used to say, "Carpenters bang their fingers, electricians get shocked, and firemen get fucking burned." Under Gallagher, Rescue 2's men spent their share of time in the hospital, but, miraculously, no one got killed.

With Gallagher sending his guys into every big blaze in Brooklyn, Rescue 2 fought so many fires that they developed their own new fire-fighting strategies, too risky to put in the department procedures but not too risky for Rescue 2. One of these strategies had to do with row houses, private houses connected to one another, which lined the streets of Brooklyn. Typically, all the houses on the block shared a cockloft, the area between the top floor's ceiling and the roof. No walls separated the cockloft of one house from the next. This meant that fire could shoot down the block in no time and take down every house. So Fred Gallagher and his men came up with a strategy for fighting row house fires. They'd get to the last house that wasn't burned down and run up to the top-floor kitchen. Then they'd break a hole in the ceiling and boost Jack or another short guy up on top of the refrigerator. Gallagher would get on the radio and tell his roof man to cut the roof above them just as Jack was hitting the fire in the cockloft with a hose line. The heat and smoke would vent through the roof, and Jack would work the nozzle, head stuck up in the ceiling. Eventually, he could crawl in and chase the fire back where it came from, saving the block.

WHEN THEY WEREN'T FIGHTING FIRES, in the hours between blazes when they wolfed down a meatball hero or a plate of pot roast, Gallagher's Rescue 2 built another tradition: relentless abuse and provocation in the kitchen. Through all of the mayhem and chaos of the time, the boys had fun. They played practical jokes on each other, told tall tales, and teased new Rescue 2 men who didn't yet know how to defend themselves against the onslaught of abuse. No one could stop Jack Pritchard from busting balls. Captain Gallagher had a bunch of characters on his hands, insane, eccentric guys who were united by their love of the job and their love of a well-executed jab or insult.

A lot of these men were resolute individuals, but the captain worked hard to break them down into a team, which made them stronger than just six men working separately. Gallagher wouldn't tolerate lone rangers, and if his words couldn't persuade, he let his brawn do the talking. He could take on five firemen even if they all attacked him at once.

Jack once had the bright idea of ambushing Gallagher when he entered the kitchen, and enlisted the help of the other firemen on duty. John Thomas, another tiny but powerful hellion, went right for Gallagher's legs, where he had just been badly burned. Gallagher punched him in the ribs and cracked three, disabling John. In rapid succession Gallagher took down three more guys. Jack, the only man left, yelled, "I give up!"

"You can't give up, Jack," said Gallagher. "You're a fucking dead man." He picked up Jack, flipped him upside down, and pile-drove him into the concrete floor. It was tough love, Fire Department style.

But Gallagher could also persuade by gentler means. When the house had to be painted before an inspection, he woke up at 3 AM and started painting the bunk room where the men slept. Before long, the men had climbed out of bed and grabbed brushes and paint. Once the paint job was under way, the captain snuck back to bed and fell asleep.

Gallagher had opened an ill-fated business to supplement his captain's wages: a candy store near his home on Long Island. The store didn't last long before it had to close; that just proved what everyone knew—Fred was made to be a fire captain, not a candy-store clerk. Once, when they had a fire in Williamsburg, Gallagher had Tom stop off at a knish place so he could pick up two dozen to sell in his candy store. Later that night, just before dinner, Jack stood up in the kitchen and announced, "We got nosh." Then he opened the oven and took out twenty-four piping hot knishes that he had stolen from Gallagher's car, which he had opened with a slim jim. Again the bear roared, "Jack, you're a fucking dead man." And again he nearly killed the guy.

Jack was always pushing his luck. When one of the lieutenants went on vacation, a lieutenant from another company, who moonlighted as a bouncer, replaced him. Even Jack didn't dare fuck with a professional bouncer. But, on the lieutenant's last night, he couldn't resist, especially after the guy mentioned that his wife had baked "two beautiful cheesecakes for my guys on the day tour tomorrow."

When the lieutenant went upstairs, the firemen said to one another, "Why the hell is he telling *us* about the fucking cheesecakes if he's not giving us any?" Jack just couldn't resist breaking into the guy's car and boosting the cakes.

He waited until they were all seated around the dinner table, diving into plates of spaghetti and clam sauce, before he announced, "We got dessert." As the lieutenant looked up from the opposite end of their rectangular table, Pritchard opened up two bakery boxes to proudly unveil the lieutenant's cheesecakes. The bouncer took off, climbed on top of the table, and sprinted across everyone's plates of spaghetti, ready to strangle Jack. But Jack saw him coming, cocked a plate of spaghetti and clam sauce, and tossed it in the bouncer's face. The lieutenant slipped on another plate and slid right off the table, landing on the floor. The men took hold of the cheesecakes and jubilantly smashed them over his head.

IN THE SECOND HALF of the seventies, Gallagher's Rescue 2 distinguished itself as the company that rescued other firemen and civilians when things went really bad. Three incidents between 1976 and 1978 came to define that era.

The first came at 6 AM on January 7, 1976. An engine company and a truck company were working inside an A&P Supermarket on Atlantic Avenue and Clinton Street, near Brooklyn Heights, when the first floor of the building collapsed. Seven firemen plunged into the basement. When they landed they were at the bottom of a pit filled with debris, screaming for help in the dark while the fire was free burning around them.

Pete Bondy was working the A&P fire without Captain Gallagher, Jack, or Tom. At first, after the floor collapsed and these seven guys fell in, he tried to climb down a ladder directly into the basement, where the men were trapped. But it was just too hot to make it down there on the ladder. So Pete and some other firemen tried to get into the basement from another entrance. But that proved impossible too because, when the big A&P had been created out of a few smaller stores, construction workers had knocked down all the partitioning walls on the first floor but left them standing in the basement. The men were trapped in a section of the basement that didn't connect to the rest. The only way to get to them was from above or by breaking through a wall from one of the adjoining rooms in the cellar.

Pete joined a crew of men down in an adjoining part of the basement, swinging a battering ram to breach the wall into the area where the men were trapped. A team of three firefighters would take ten swings, then hand off the battering ram to another team. After hundreds of swings they finally burst through the wall and opened a hole to the area where the guys were. A lieutenant tried to climb in but got disoriented when the floor dropped down precipitously inside the hole. He came back out through the hole smoking. Pete organized three other firemen to follow him in there and find the missing members. Pete went first, following the sounds of voices. He sifted through the debris with his hands until his glove grabbed a boot or a helmet. Three times, he found a fireman who was still alive. Three times, he rapidly tied a rope around the victim and, with the help of other firemen, dragged him out of the smoldering basement.

One hour after the collapse, five of the seven men had gotten out. Three had come out with Pete, and two others had been able to climb up a ladder to safety. Two men were still inside the burning collapse. Pete wanted to go down once more, but he was shot. He could hardly climb up the stairs to get fresh air. But reinforcements had arrived: Fred Gallagher and Tom Dillon. They had heard the news when they arrived at Rescue 2 that morning and rushed to the scene immediately.

Now Gallagher took charge, heading straight for the basement and entering the collapse where Pete had helped make the hole. It got so smoky and hot down there that Gallagher told Tom and Jack Carney, another Rescue 2 guy, to back out and protect him with the hose line. Rummaging in the dark, Gallagher's glove touched a body. The guy's mask was still on his face, but his air supply was completely depleted. Rescue 2 had gotten there too late for this fireman. But Gallagher didn't tell the chief that the man was dead. If he admitted that to the chief, the chief might pull Gallagher out, and he wasn't leaving until all the guys were accounted for. There was one more fireman in there, and he might still be alive. So Gallagher simply told the chief that he had found a guy and was working on getting him out. A mobile hospital with the Fire Department's chief surgeon was standing by next to the collapse. After the victim, Charlie Sanchez, was finally dragged out, they rushed him into the trailer and tried in vain to resuscitate him.

Gallagher and the other two firemen were already back in the basement, hunting for the last man in the burning debris. Suddenly an enormous air conditioner came crashing down on the basement and the fire lit up. It had been smoldering before, but now the air from the roof was feeding the flames. It was really ripping now, way too hot to do any exploring. The chief ordered Gallagher and his men out. But Gallagher, who knew the fire would cool down after it had vented a little, wanted to stay. He hesitated. Now the chief was screaming at him to get out.

Gallagher stomped out of the hole, into the adjoining room in the basement. "We gotta find that guy," he insisted. But the chief on scene told him he couldn't go back in. Gallagher protested. The chief refused to budge. If Gallagher wanted to argue, the chief said, he should take it to the chief of department.

So Gallagher hustled up the stairs to the street and went straight to Chief of Department John O'Hagan. "Fred, I can't let you go in there," O'Hagan told him. "You'll get killed." Gallagher casually walked away, ducked back into the basement, and went to the chief on scene.

"O'Hagan says we can go back in," Gallagher told the chief. Immediately Gallagher and the other Rescue 2 men climbed back into the hole. Now the search was more urgent. A lot of time had passed. But luckily the fire had cooled off a little, having vented after the air conditioner's collapse, and the men were able to do a more thorough search. They lifted up a shopping cart and touched a uniform. As they uncovered the victim's body, he suddenly yelled, "Get me the hell out of here!" The guy was kicking and screaming so much, trying to get out, that they had a hard time strapping him into the Stokes Basket.

As soon as he exited the building, Gallagher fainted, overcome by exhaustion and smoke. He was rushed to the hospital along with Tom, Pete, and almost every other man from Rescue 2. But by the time they arrived at the hospital, all the guys were feeling better. At the hospital, Tom Dillon even ducked into the bathroom for a cigarette. He had just lit up when the bathroom door opened wide, and in stepped the FDNY's chief medical officer. "What the hell are you doing smoking?" the guy screamed. "You're supposed to be in for smoke inhalation. Get outta the hospital!"

He had blatantly disobeyed his chief of department, but, nevertheless, a few months later, Fred Gallagher walked up to a podium in front of city hall to accept a medal for his disobedience. He won the Gordon Bennett, the highest department citation, for his rescue at the A&P Fire. And Pete Bondy and Jack Carney also got medals for that job, where they had helped to rescue five out of the seven firemen who had been caught in the collapse. It was jobs like the A&P Fire that helped build the Rescue 2 legend.

One of the other nights that defined the era—both for Rescue 2 and for all the people of New York—was July 13, 1977, the night the lights went out. Just after the power outage, the city seemed like it might remain calm, as it had during the blackout of 1965. But over the intervening twelve years New York had become a very different place, and, before long, the looting started, and the ghettoes of Brooklyn began to burn.

The Rescue 2 men were raring to go when their first assignment of the night came in: Coney Island, where people were trapped in a stalled amusement park ride. The guys bitched and moaned all the way there. They wanted to go to a fucking fire. There had to be twenty fires burning in the borough, and they were heading to the amusement park to play around with a ride. Unbelievable.

By the time Rescue 2 finished at the roller coaster, the dispatcher was summoning them to Maimonides Hospital, where hundreds of injured people were left helpless when auxiliary generators in the OR and ER failed. The looting and rioting had caused terrible injuries—including people who went through plate-glass windows and severed arteries. But doctors couldn't help them much without light and monitors.

The Rescue 2 guys rigged up an improvisational OR and ER in the Maimonides parking lot, lit by lights and a gas generator from the rescue rig. Even as they set up their lights, more patients were arriving every few minutes. By midnight Rescue 2 was gone, leaving the doctors with a night's supply of gas before speeding off to fight some fires.

By this time there were so many buildings burning in Brooklyn that the dispatchers could only send units to occupied buildings; vacants were left to burn to the ground. But the dispatchers knew that firemen couldn't resist stopping at the first good blaze, so they had to get on the radio and say, *"You're gonna pass a six-story building on your left that's fully involved. Don't stop."* The Rescue 2 men sped by fires that on other nights would have merited multiple alarms. They went to fifteen fires that night and drove by dozens more. When the sun started to rise over Brooklyn, the men could look out the back of the rig and see the shopkeepers sitting out in front of their stores, shotguns across their folded legs, guarding the businesses they had invested their lives in. Driving the rig back to the firehouse after the last run of the night tour, Tom Dillon saw a woman struggling to carry a table she had looted. It was beat-up and broken with only three legs left, but she was determined to schlep it home.

FRED GALLAGHER LED HIS MEN into every kind of fire, and he always got them out. Under his leadership, Rescue 2 became the company other Brooklyn firehouses reviled in public, but admired in private. The word around Brooklyn was that Gallagher's men would do anything to steal the hose line, including turning off a fireman's air supply, knocking guys down, or even pushing them into the fire. All this was rumor and nothing more. Part of what fueled the rumors was the hard personalities at Rescue 2: these were men who weren't afraid to shout out "coward" to other firemen in the middle of a blaze, something that almost always brought men to blows. Such behavior didn't win the Rescue 2 guys any friends, but it did get attention, along with the stories of their heroics at fires. And those stories never stopped coming.

It was 8:55 AM on August 2, 1978. Jack Pritchard had been driving the night before, and he had five minutes left before he could go home. When the alarm bell sounded, the day-shift driver said to him, "I'll take it in."

"Fuck you will," Jack replied and jumped into the driver's seat on the rig. The call was for a fire in a Waldbaum's supermarket—another deathtrap. When Rescue 2 arrived, the other companies were fanned out all over the building, with men up on the roof, on the supermarket floor, and on the mezzanine level, where an office had been constructed.

Phil Ruvolo's cousin and best friend, Bill O'Connor, was up on the roof. Standing on the street below, Bill's wife and kids waved up at him. In those days many firemen worked in their own neighborhoods, and their relatives sometimes came to cheer them on. Bill had just started on the job, and his wife and kids had come to see him at work before a day trip to the beach. This fire was near Phil Ruvolo's home in Flatlands.

Jack was following some other firemen up the stairs to the mezzanine when he got the feeling that the fire had spread more than anyone realized. So Jack ran down to the supermarket floor to punch a hole in the ceiling and look for fire. As soon as he punched the hole, the roof and the walls came down on top of him. As he went down, Jack saw Tommy Valbuona, Rescue 2's roof man, flying down through a hole in the tin ceiling. Tommy's head was swollen to twice its normal size from severe burns. Tommy fell through the hole and landed near Jack. Jack grabbed him and said, "Show me where the other guys are." Tommy pointed to some aisles that were now shrouded in the collapsed ceiling. Jack told him, "Get out of here."

It was Rescue 2's job to go for the missing firemen. Jack went looking and quickly found a victim. He hauled him out to safety and went right back in. After Jack's first rescue there was another collapse. More of the roof came down, making it even harder to search. But somehow Jack managed to uncover a second fireman in the burning debris and drag him out too. After this the whole ceiling came down, but Jack clambered on top of the pile and dug out one more fireman.

After that third grab, Jack went in for a final search, crisscrossing every inch of the supermarket floor. If anybody was alive in there, they were fading fast. Jack yelled out at the top of his lungs. He went back and forth, making sure he covered every aisle in the place.

Suddenly he stopped dead in his tracks. Where was the exit? He'd been in there three times already, but now the place looked different. He couldn't find the way out, and he was out of air. Calm down, he told himself. He took a close look at where he was and finally got his bearings. He struggled toward the exit and out to the sidewalk, where he told the chief on scene, "There's no one in there. They're all out."

"They were on the roof, Jack," the chief replied. It was like a kick in the gut. The rest of the missing men had been up above, where Jack couldn't get to them, and that meant they were dead, burned beyond recognition. A lot of guys were up there. Bill O'Connor, who had waved to his wife and kids, was up there, and now he was dead too. It ate at Jack to know that there was no way he could have saved those men. He was in Rescue 2. His job was to get every man out.

Meanwhile a young Phil Ruvolo had woken up to find his cousin Bill's wife at the door, telling him to come quickly to the Waldbaum's. When Ruvolo got there, the firefighters were still trying to find his cousin Bill, but from the look of it Ruvolo thought there was no way he was still alive.

By THE TIME Fred Gallagher left Rescue 2, in 1980, it was the hardest-working unit in Brooklyn. The Rescue 2 that Phil Ruvolo would come to command in 1998 was forged in the seventies. The ballbusting attitude that was kept alive by men like Lincoln, Richie, and Bob started with guys like Jack Pritchard and Fred Gallagher. In his first months at Rescue 2, Ruvolo would hear the Gallagher stories told in the kitchen by the senior men, who had heard them years ago and passed them on.

It was no ordinary company that Ruvolo was leading. It was the elite of the FDNY, the guys with the best skills, the most pride, the toughest attitudes, the greatest willingness to risk their own hides—and some ineffable extra quality that could only be described as Rescue 2, Brooklyn.

All the stories, legends, and lore told Ruvolo just how much he had to live up to. In his early days, the Rescue 2 legacy made him even more determined to put his own stamp on the firehouse. Occasionally, he runs into the legendary characters who created his company, guys now going gray or gone white, thinned or rounded out from old age. Fred Gallagher fought a battle with cancer that was as hard as any fire he ever went up against in the old days. But Ruvolo can see the glint in his eyes that explains all the stories about Gallagher and his men that still get told at Rescue 2.

# FOUR: ATLANTIC AVENUE

**C**APTAIN PHIL RUVOLO ENTERS his office on this morning, April 14, 1998, and unwraps a laundry-fresh work shirt, crisp and clean, stiff with starch. Today he'll be getting his first new man in the company, a kid named John Napolitano, who comes from a Queens ladder company and has spent the last few months driving a chief. Ruvolo, not exactly excited by this prospect, anticipates some fireworks. Life for newcomers is never easy at Rescue 2. This is a house where a guy who has been around for five years of service is described as having just stopped by for "a cup of coffee." One of the firemen who left after serving two and a half years at Rescue 2 was famously awarded a plaque with half a cup of coffee on it. John, however, should expect no plaques in the immediate future. These guys won't give him a cup of anything. There are a couple of reasons for their animosity. First, he got his spot at Rescue 2 because, as the guys put it, he was "hung"; that is, he had political weight—and got a little help from the chief he'd been driving. Even though half the guys in the firehouse got here with a little help from their friends, that won't stop them from brutalizing John for using his influence.

Moreover John has already made a serious breach of etiquette. At Rescue 2, guys may fart publicly with impunity, talk with their mouths full of chopped sirloin, and address one another with terms of endearment like "fuckhead" or "pig," but they do have their own code of conduct, as veiled as the Freemasons' and as complicated as Scientology. According to Rescue 2's closely observed yet entirely unwritten strictures, John should have made a courtesy call to the new captain and introduced himself before coming to work. He didn't.

Besides these offenses, John has one even bigger demerit against him: He's not a product of a Brooklyn ghetto firehouse; he's from a slow company in a borough that most Rescue 2 guys consider the equivalent of a volunteer company in Iowa. The guys will excuse any offense committed by a man who really knows how to fight fires. But, to them, John's pedigree isn't promising.

Rescue 2 draws most of its new recruits from nearby Brooklyn ghetto companies, all of which, like Rescue 2, look down on the pretty boys in Manhattan and the guys who twiddle their thumbs all day in Queens. Among firemen, there is a complicated hierarchy of companies based on who goes to the most real fires in the city. "Run statistics," which show how many times a unit goes out the door and how many fires it responds to, don't always tell the whole story. Some units are constantly out running to nothing: first-aid calls, food on the stove, mattress fires; bush-league stuff. By most people's reckoning Rescue 2 is the busiest rescue company, and the busiest truck company is either 111 Truck, just around the corner, or 157 in Flatbush.

Most Brooklyn fires are concentrated in a small chunk of neighborhoods that border one another: Bedford-Stuyvesant, Bushwick, Brownsville, East New York, Flatbush, and Crown Heights. This constellation of neighborhoods has been the center of Brooklyn firefighting for years. Time at a company in any of these areas is a pedigree worthy of Rescue 2. To come from one of the lesser Brooklyn companies is barely acceptable, but to come from Queens, with little time on the job, as John Napolitano does, is simply wrong. The stain of this original sin is going to take a long time and a lot of hard work to wash off.

RUVOLO HEARS THE INTERCOM BUZZ and then the words *"Cap, we got the new guy down here,"* muttered with minimal enthusiasm. *Let him swim with the sharks for a little while before I grab him,* Ruvolo thinks. Nobody down in the kitchen is happy about Napolitano coming here. Richie Evers, not surprisingly, has already sworn to give the kid the cold shoulder: he's planning on giving John the complete silent treatment. Others, including many of the veterans, have decided to follow suit. Only Bob has some sympathy. He knows John from volunteer firefighting in Lakeland, Long Island, where they spend their off-hours as suburban "volleys".

But even Bob can't stand up for John until he's seen him work a few Brooklyn fires. That has to come first. Like every other new firefighter at a rescue company, John is here on a six-month trial basis. If he swims, Ruvolo will keep him. If he sinks, it's back across the Pulaski Bridge to Queens.

When Ruvolo hears a knock on his door, he shouts, "C'mon in," and John enters. He is not an impressive sight: In a house where two hundred pounds is a respectable weight after three months on the Atkins diet, John Napolitano appears emaciated by comparison. To the guys he's skinny, but John hates when people call him that. "I'm athletic," he always tells his wife, "not skinny." To top it off, he looks about twenty-four years old. The only thing that betrays his twenty-nine years of age is a thinning and receding hairline. He has that eager-beaver look. A total probie.

"Sit down," Ruvolo commands, and John complies without saying a word. He has the shocked, traumatized look of a newcomer who has just braved his first few minutes in the Rescue 2 kitchen. Ruvolo, a little put out by the guy's sins of omission, is in no mood to ease his pain, and, in the ensuing—almost entirely one-way—exchange that follows, the captain demonstrates not only his ability to dominate the conversation but also his mastery of the pregnant pause and the painful silence. After thirty agonizing seconds of gazing at the new recruit with an expression of total revulsion, he sums up his thoughts. "You wouldn't have been my choice to come here," Ruvolo begins, pausing once more to inflict further torture as John fidgets in his chair, keeping his alarmed eyes fixed on Ruvolo.

"You're gonna do everything that I tell you to do," Ruvolo continues, determined to make him or break him. "You're in my group. And you're working straight tours until further notice." This means every single tour with Ruvolo and no twenty-four-hour shifts. Instead of coming to work three times a week for double shifts, John will have to come five times a week for single day or night tours only. This ensures that every time John sets foot in the firehouse, he'll be working under Ruvolo. This means that maybe the captain can keep him alive while he finds out if he has the skills to make it here. It also means that, given John's commute, he'll have hardly a second of free time in the months to come.

Ruvolo is tough, but he's nothing compared to the animals downstairs. When John steps back into the kitchen, an uncomfortable silence looms. Like Ruvolo, the men have mastered the subtleties of silence. But the chill is much worse when it's coming from four semi-dangerous-looking guys just staring at you. John knows then that earning a place in this company will be a battle.

John, however, looks willing to brave the lions. He seems ready to withstand any amount of pain and suffering to make it here. Rescue 2 is where he has dreamed of working since he first heard FDNY vets telling stories at the Lakeland Volunteer firehouse. All the suburban volunteers liked to listen to tales of disaster and mayhem in the city. But John was different. He wanted to live these tales, not just listen to them.

Junior rescue men like John Napolitano spend their first few months trying to learn everything they can while also trying to lay low, avoid trouble, and occasionally impress the more senior guys with their devotion to even the most tedious firehouse tasks. One minute a new guy may be getting instructions in rope rescue or on how to cut steel with a plasma cutter. The next minute he may be ordered to bring up a rusty tool from the basement that hasn't been used in decades and to scrub, sand, and polish it until it shines.

During his early weeks at Rescue 2, John has to appear eager to perform all the house's grimiest, most humiliating chores. But he knows he has something to make up for, and he's game: He's the first to jump on a pile of dirty dishes after meals and must carry out all the chairs and sweep and mop the floor after dinner. He empties the dish-washer as soon as each cycle is complete, and makes sure all the trash cans are empty. All this is just the beginning. A nervous new guy like John must have all the domestic intensity of a new bride on her first day at home after the honeymoon, without a second of delay, hesita-tion, or rebellion—or a scintilla of sympathy.

After days like this, John comes home from work exhausted, too tired even to move from the couch into bed. His wife, Ann, is preg-nant with their second child. Since their baby is on the way, he wants her to stay calm. So he just takes whatever the other men dish out, without divulging the humiliations. If she knew how badly the guys were abusing him, she'd be enraged.

In his locker John puts up a pretty picture of Ann and their daughter. Whenever he sleeps over at the firehouse on a night shift, he makes two good-night calls: the first to his daughter, and then, later on, a second to Ann. The guys don't tease him about this. Even at Res-cue 2, some things in a man's life are off-limits.

JOHN WAS BORN in Williamsburg, Brooklyn, in the same house where his father grew up. Even though he left Brooklyn at a young age, John was raised on his dad's stories of coming-of-age in the old neighborhood, now a hipster paradise of tattooed skateboarders, back when it was an Italian and Polish stronghold. Even then Williamsburg had its share of crime and drugs. The Northside of Williamsburg was the kind of place where kids would skip over junkies and dodge book-ies on their way to school. In his youth, John's father used to play on the sidewalk, kicking cans that he knew concealed betting slips for a local bookmaker. When he got home his dad said, "C'mere. Why'd you do that?"

"What?"

"You know what. Don't do that again."

Kids who grew up on John's father's street became cops or gang-sters. Some became both. When John senior moved his family out to Ronkonkoma, John was just five years old. His dad felt like a pioneer; across the street from the new house was nothing but woods. Next door was a horse stable. He watched his son play on a real lawn and run off into the woods. The schools were better than the ones in Brooklyn, and John's dad had big plans for his son: he always dreamed he'd become a doctor one day.

But when he was seventeen years old, John came home with a fireman's uniform. At first his dad thought it was a Halloween costume. It wasn't. It was the first sign of John's newfound passion. Not long after, John signed up for EMT courses and started riding in the ambulance as a junior paramedic. Other kids hung out at the bars, but John would sit around the local firehouse, listening to tall tales from guys like Bob who worked as firemen or cops in the city. At that firehouse, John had found his place. He loved the fireman's job so much that he wanted to do it for a living. But getting into the FDNY was hard and he had to cover all the bases. So John took the tests for what seemed like every civil service job on the eastern seaboard—city cop, Suffolk County cop, fireman, corrections officer. After the usual anxious wait for the test results, he passed and, as soon as an opening came up, he joined the FDNY.

John's first day at the Rock, the fire academy, was pure hell. While he was running, climbing stairs, trying on gear, he got horrible stomach pains. *Maybe I'm not cut out for this,* John thought. He came home after a horrible day at the academy and found out that his dad had accidently poisoned him the night before—everyone was sick from his lasagna.

Right after John started working for the Fire Department, he got a job offer from the Suffolk County Police. It was one of the best-paid civil service jobs around and he'd be working right near his home.

"Why don't you take a leave of absence from the Fire Department and try it out?" his dad suggested.

"I love my job," John said, and that was the end of the discussion.

At his first firehouse John was a practical joker. He would short-sheet beds, frustrating men trying to get to sleep. He'd saran-wrap the toilet seat or rig the faucet so it sprayed you in the face. At Rescue 2, however, John is a different man, dead serious almost all the time, too afraid to crack a smile.

After his first tense moments at Rescue 2, John decides that his strategy is to work hard and humbly. Perhaps by the onset of his middle age or so, the men will come to accept him and he will again feel like a real member of the department. Right now he's just trying to survive. During the big, laughing, rollicking discussions in the kitchen, he stays out of the fray and makes no effort to command attention. For the first couple of weeks, he hardly says a word. He watches, listens, and attempts to learn. Usually, however, he is either cleaning pots or scrubbing tools. He studies every piece of equipment, every exchange at the kitchen table, as if he were boning up for an exam. He tries to figure out how the place works and how he can get these guys to stop hating his guts.

AFTER FIREMEN get through this grueling initiation period, they get a lot more liberty to do what they want whenever they want. Everybody has a favorite place in the firehouse to kill time. Bob, of course, can always be found in the kitchen, hovering over the stovetop, cooking or cleaning. Ruvolo drifts between the kitchen and his office, wandering down periodically for one of the eight to ten cups of coffee he consumes each tour. He tries to spend as much time as possible down with the men, but a deluge of paperwork often keeps him upstairs. Lincoln hangs around in the shop working with the torches. Periodically, he peeks into the kitchen to see if he can start any trouble. (By now, the guys are wise to Lincoln's talents as a pot stirrer, and they resist his attempts to get them riled at one another. But that doesn't mean he's stopped trying to have some fun. Every once in a while he still succeeds in starting a scene worth watching.)

Rescue 2's firehouse is small and bare by FDNY standards. Most other firehouses have at least a den or TV room and usually a weight room. Some of the houses in Manhattan even have basketball or handball courts. But Rescue 2's Bergen Street home is as basic as it gets. The front door opens into a garage just wide enough for the rig, and the concrete floor is speckled with oil leaked from the truck. To soak up the spills, the men dip into a barrel of sawdust and toss it on the floor.

Up near the front door there's a big map of the borough tacked to the wall—for those rare occasions when the chauffeur needs extra guidance. On the opposite side, a snaking plastic pipe hangs from the ceiling. When the truck comes in, someone clamps this pipe onto the exhaust so the guys inside Rescue 2 don't suffocate from the fumes.

At the end of the apparatus floor, the garagelike space where the rig gets parked, are a tiny bathroom and the company workshop. Behind the floor is the kitchen, cavelike and ancient, which looks like a relic from some Stone Age tribe that lit cooking fires indoors. Black streaks mark the wall and ceiling above the stove, where Paul Somin once left a pot cooking when the company went out for a run. They returned hours later to find the first and second floors thick with smoke, and had to make a push into their own kitchen to extinguish the fire. The kitchen is a fixture on the citywide list for renovation. The men have waited years for improvements—don't raise the subject with Bob—so long, in fact, that no one knows if they could actually accept a change. The city's complex method of open bidding, designed to save money and ensure fair play, requires so much bureaucracy and paperwork that it invariably saves money because nothing ever gets done.

In one corner of the kitchen, next to the drinking fountain, is a computer terminal. The computer has replaced the department's old Morse code system, which was once used to summon companies. Back in those low-tech times, one fireman was always designated the house watch, and his main task was to transcribe the messages and ring the bells when his company was summoned. Now, when an alarm comes in, the computer does all the work, and the nearest guy just hits a button on the screen to acknowledge the run and tears off a printout from the dot matrix printer. In another corner of the kitchen are a big TV and four couches arranged in an L-shape along the walls. There's a small area out behind the kitchen with a barbecue grill where the men char-grill steaks in the summer. The tiny backyard is surrounded by thirty-foot-high fencing, along with thick plywood that goes from the ground up about twelve feet, to keep out thieves who used to clip the fence and break in.

Downstairs, the basement is a dirty, dusty dungeon with a few old weights lying next to a weight bench and a decades-old boiler that, every winter, struggles to heat the place. Upstairs is the locker room, filled with big metal lockers and more smoky, sweaty old clothes and gear. When Rescue 2 fireman Ray Smith came to the company a few years ago, he got accused of using more than his one designated locker. He had secretly commandeered another empty one and was warehousing his junk there. Somebody then went around the firehouse and stenciled "Property of Ray Smith" on every locker in the firehouse. Ray got the message.

Off the locker room, there's a small room for the officer with its own private bathroom, and in the rear there's a bunk room with six beds and a department radio, never turned off, so the men can hear what's happening in the borough of fires even when they're drifting off to sleep. On top of one bunk is a pile of clean sheets and blankets; when a fireman finishes his tour in the morning, he tears off his old bedclothes and replaces them with these. Though the place isn't the Waldorf Astoria, the guys are fastidious about the basics like changing the sheets, cleaning the bathrooms, and emptying the trash cans. Many visiting wives have wondered aloud on their tour of the firehouse how the same men who keep this place spic and span can be such total slobs at home.

Usually, only the junior man sleeps in the bunk room. The other four firemen on duty during each shift prefer to sleep downstairs. A guy who lives too far away to commute between shifts might also sleep in the bunk room. When the meal is ready, or when there is a promising job on the radio, the guys downstairs ring the bells to signal everyone to head down. One bell means a phone call for the officer. A couple of rings mean that chow's on. Quick, frantic bells mean a run.

The officer always bunks in his room. Though the officer joins the men in the kitchen for all the roughhousing, fun, and stories, there are some rules that are never broken: The captain never cooks or cleans. He carts only one small Halligan tool into a fire, never a big hook or a can. He never answers the telephone in quarters. And the men don't ever call him by his first name; at work he's Cap, Cappy, or Captain, even to his closest friends.

At the top of the stairs is a bulletin board, with FDNY and Rescue 2 notices. The biggest sign gently reminds men who still haven't paid this quarter's house dues: "PAY THE FUCKING HOUSE TAX, YOU BUMS!" House taxes are what the men pay out of their own pockets for items like coffee, milk, butter, salt. The rest of the notices tell of various meetings, memorial runs, and fund-raising dinners for groups like the Emerald Society, composed of Irish-American firemen, and the Holy Name Society, a Catholic organization. Then there are the T-shirts, concert tickets, and other goods put up for sale by enterprising firemen. With a list of a few hundred firehouse fax numbers, a guy can get the word out to almost eight thousand men in one morning.

Finally, at the end of the bulletin board, there is a big glossy advertisement for supplemental life insurance, promising to help out a fireman's family if he gets hurt or killed on the job. One of the insurance salesmen was here recently, glad-handing in the kitchen and spelling out the details of the payouts if a guy gets hurt or killed. Some men plan meticulously for their future. Lincoln and his wife have already discussed what she should do if, God forbid, he gets killed. Others feel that even discussing insurance payouts or funerals only makes those possibilities more likely.

Each corner of the firehouse embodies the past and the future of the place. In the area where Lincoln works on the torch, there is a row of metal spikes. These spikes have been extracted from the bodies of impalement victims, all saved by Rescue 2. The company practices on these gruesome trophies to learn how to cut different materials with their tools. When somebody falls on a sharp metal fence, the paramedics can't just yank the victim off. They need Rescue 2 to cut the fence around the injured person and bring him or her to the hospital with the metal in place—otherwise the person might bleed to death from the open wound. Sometimes the guys even follow the victim into the OR with their torches, in case the doctors need them to make more cuts.

Throughout the firehouse, the walls are covered with pictures, plaques, and medals, marking the milestones of Rescue 2's seven decades. At the bottom of the stairs hangs a picture of the company's first, now ancient, rig from 1925. It looks like an old milk truck, and the firefighters are arranged at the side of the truck with twenty-five identical, earnest smiles on their faces. Next to this photo is a wall covered with unit citations—certificates that the company receives for distinguished action (awarded when a chief notices them risking their hides). To Ruvolo these citations don't mean much. What impresses a chief, looking in from the outside, is rarely the same as what impresses firemen. Some guys call the display of certificates "the wall of lies." Medals and all the rest come when a fireman happens to be lucky enough to stumble over an injured person and drag him out. Far more important to firemen is putting the damned fire out. That saves everyone but it's rarely the reason you get a medal.

Just as the walls hold memories, grouped by decade, so do the current firemen divide neatly into generations. At the top of the heap are the Gallagher guys, who have seen action since the seventies. Now, almost twenty years since Fred Gallagher's departure, the only man of this generation still in Rescue 2 is Richie Evers. Next are the guys from the days of Captain Downey, including Dave Van Vorst, Timmy Higgins, Billy Lake, and Bob Galione. Finally, there are Captain Fischler's picks: men like Kevin O'Rourke, Paul Somin, and Lincoln Quappe. Though there is definitely a firehouse type, personified by the hardworking, hard-talking, take-no-bullshit attitude of Richie Evers, a lot of different personalities have managed to squeeze under this roof over the years. But up until now, there's been no room at all for skinny boys from firehouses in Queens, guys like John Napolitano.

IF THERE IS A RAY OF HOPE for John as he polishes the tools and scrubs the kitchen counters and endures the silent treatment, it is the fact that Kevin O'Rourke holds a place of honor in this company. It was inevitable that Kevin would make a strong impression on John, because for his first few weeks at Rescue 2, Kevin O'Rourke's is one of the only friendly faces in a sea of hostile scowls.

Kevin O'Rourke is almost the exact opposite of the Rescue 2

stereotype: he's unfailingly compassionate, he seems incapable of complaining, and the worst thing he's ever called somebody is a "poopy-head." The one thing he has in common with the other guys is his firefighting ability. Ruvolo thinks of him as a fireman he's glad to have next to him when he's trying to make one last push against a blaze.

At first glance, Kevin is way too nice to fit in at Rescue 2. He's the kind of guy who met his wife in their church's marching band. (They both played trumpet.) Even the story of why he became a fireman is heartwarming: it dates back to the time a firefighter carried him out of a fire in the family apartment when he was eight years old.

Kevin got his first taste of the firefighting life while still in the marching band. When they went to play at parades, he never passed up an opportunity to hop on a fire truck and check it out, or to chat with men he yearned to emulate. As soon as he was old enough, Kevin joined the volunteer fire department in his hometown. As soon as he passed the fireman's test, he joined the FDNY.

When Kevin arrived at Rescue 2, he threw the guys off at first because he was so relentlessly polite. When a kid came knocking on the firehouse door for air for his bike tires, Kevin wouldn't just lead him to the pump; he'd give the kid's bike a ten-point checkup. He'd adjust the seat, calibrate the brakes, and make sure the bike rode smoothly.

By fire department standards, he was also enormously considerate of his wife, Maryann. The first time he got hurt fighting a fire, he snuck out of the ambulance and trekked over to a nearby bodega with his oxygen tank in tow so he could use the payphone. Sheepishly, he called his wife. "Nothin' much goin' on," he said. "I got a little hurt. Don't worry. I took a little smoke and they wanna take me to the hospital for a few routine tests. Nothing to worry about. Mare, you're not mad, are you?"

"No, Kev, I'm not mad."

By the time Kevin had trudged back, the paramedic, thinking he'd misplaced his victim, had raised a chief. "Sit your ass down there and don't move," the chief told Kevin. But Kevin was happy—at least Maryann had heard his voice and knew that he was all right. If someone called her and told her he was hurt, who knows what she would think?

When Kevin comes home from work he sometimes share stories with his wife. But he saves his death-defying tales for his older daughter, Corinne, a teenager who loves to hear about her Dad fighting fires. His younger daughter, Jamie, still in middle school, lacks Corrine's distance. She always comes close to crying when she thinks of her father in danger. Everything about firefighting strikes terror into her heart.

On one Christmas Eve, when the girls were younger, they were on their way home from a movie with their dad. A fire had just been reported nearby, and Kevin whisked the girls over to the fire, stowed them safely with another fireman's child in a chief's car, and then went in to fight the blaze. When the windows in the involved building shattered, the boy told Kevin's daughters, "Your father probably just died in there." His younger daughter, Jamie was inconsolable and hysterical. Corinne, the older daughter, had to find a fireman and ask him to bring out their daddy so her sister could see that he was still alive. When they went home, Kevin hurried upstairs immediately for a quick shower to wash off some of the smoke before attempting to explain to Maryann why Jamie was sobbing uncontrollably.

"Kev, why did you take them to a fire?" Maryann asked after Kevin had come downstairs, still reeking of smoke.

"What did you want me to do?" he asked. "They were there with me."

Maryann had to laugh. He was such a fireman. "How 'bout not go. Did that occur to you, hon?"

The tough old bastards who ruled the Rescue 2 fortress had never seen anyone as nice as Kevin O'Rourke. They were taken aback. There was impassioned talk in the kitchen about trying to bring him over to the dark side. But the guys couldn't succeed in channeling Kevin's boundless energy into anything negative, violent, or sarcastic, the directions they were looking to pull him in. He would much rather be sweeping, mopping, or cleaning than sitting around cursing out other guys in the kitchen. He took on all the little responsibilities that no one else wanted. Maybe it was the fact that he had once worked as a janitor, but if the toilet paper needed changing, Kevin had the key to the holder. And he always had a huge smile on his face, a smile radiant enough to buoy anyone's spirits, even Bob's or Richie's. Of course, sometimes the guys got pissed at Kevin for being so fucking cheerful all the time.

But what won the guys over was the way Kevin worked a job. At a fire, Kevin isn't just some tender-hearted fellow. He's single-minded in his desire to get in there and get dirty. But even when he's barreling down a hallway, trying to pass some guy from the engine, instead of "Move it motherfucker," he'll say, "Excuse me, pardon me." Most guys are so shocked by the courtesy that they actually let him by.

Kevin is also the only guy in the firehouse who is civil to buffs. Since Fred Gallagher's days, Rescue 2 has been the number one stop on the fire buff's tour of New York. Back in the late seventies a Rescue 2 fireman came up with the idea of creating Rescue 2 patches, hats, and T-shirts. Now it sometimes seems as if every volunteer firefighter in the country is knocking on the door and asking to buy a T-shirt or hat. They don't really come to buy the merchandise; they come to get a glimpse inside the famed firehouse, hoping against hope that the guys might ask them to ride along. The Rescue 2 men rarely oblige—the idea of adoring fans, adoring male fans, makes the guys a little uneasy.

There are some buffs, however, who come with an in. Guys like Sandy Williams and Warren Fuchs are virtually a part of the company, honorary members in the eyes of Rescue 2. Other guys know someone and call ahead to arrange a ride along. A couple of guys from Chicago come for a few days every year and bring a big cooler of midwestern steaks, which definitely helps to ensure a warm welcome. Another group of guys spent an entire week at Rescue 2 without chipping in for a meal. Their last night they magnanimously offered to buy dinner. When they came in with some boxes of Rice-a-Roni, Bob tossed the boxes in the trash can and booted the men out. A few of these buffs have bordered on the fanatical. One guy came from out of town and let it slip that he was staying at a swanky midtown hotel. The guys couldn't believe he was sleeping in this Spartan place when he had a comfortable room in Manhattan. Then they heard the kicker— it was his honeymoon and he'd left his new wife alone in the city while he came to visit the firehouse. "Give us the room number," someone piped up from the couch, and the guy looked embarrassed for about ten seconds, before he started asking more eager questions about the rig.

Kevin greets buffs with his usual polite good cheer. More typical of Rescue 2 is the kind of response buffs get if they cross Bob. A favorite tale around the kitchen table goes something like this: Two buffs come knocking at the Rescue 2 door one day when Bob is on duty. They resemble firemen, but their clothes are too clean and new for them to actually *be* firemen. Even if a Rescue 2 fireman puts on a brand-new T-shirt at the start of a tour, fifteen minutes in he'd have covered the front with tomato sauce, have snagged one sleeve on a sharp edge, and have ripped the seam under the arms when he bent up to reach for a piece of equipment. But not the buffs.

If these buffs had been to Rescue 2 before, they would have known to bring a layer cake or some juicy steaks. But these two buffs come empty-handed.

At first, Bob feigns interest in them. That catches them off-guard, because they're used to being verbally abused as soon as they walk in the firehouse door.

"So how'd you come up from Baltimore?" asked Bob.

"We took 95 right up to the Bayonne Bridge," said one of the buffs.

"And then how'd you come across Staten Island?" pressed Bob.

"We took 440 right to, what was it, 278?" the buff told him.

"Oh, okay, you came that way. Now from 278, what'd you do to get over here to Brooklyn?

"Yeah, we came on 278 right to the Verrazano Bridge." And on and on until the exact route, street by street, to 1472 Bergen, has been established publicly for everyone in the Rescue 2 kitchen.

Now Bob throws down the gauntlet. "You came all the way up here, on I-95, 278, the BQE, Atlantic Avenue, and you didn't even pass a fucking bakery?"

Ruvolo loved it. Bob was brutal, just like him. The kitchen erupts in deep belly laughs, and the buff's face turns as red as the fire truck.

RUVOLO RIDES JOHN HARD. He's always coming up with some new way to test the kid. As master of the rescue ropes, Ruvolo often uses John to demonstrate exotic and painful possibilities. He'll send John off the roof hanging upside down and then make the guy pause in midair, as he might have to do during a scaffold rescue. When John gets to the ground, his eyeballs are bursting from their sockets. But John doesn't complain. He rights himself, walks over to the guys who are enjoying his pain the most, and just smiles. He knows one basic rule of this firehouse: the guys are like sharks, and if they smell blood, they'll eat away at your wound until you die.

Slowly, in his quiet way, he begins making friends with some of the younger guys, like Mark Gregory, who can sympathize with him. Mark got singled out when he joined Rescue 2 because his dad, Stephen, is an FDNY assistant commissioner. Immediately, Mark was known as "Junior" in the firehouse and had to take licking after licking. Like John, Mark is ambitious and wants to move up in the ranks. When they realize that they're both studying for the lieutenant's exam, they begin to retreat upstairs during their downtime to quiz each other on the material. John doesn't slow his pace on his chores, but when he's not scrubbing or shining, he's studying. Whereas Mark casually skims the material, John is obsessively organized. He has five different-colored highlighters, and he meticulously goes through each volume of the two thousand pages of material, classifying each piece of knowledge with a neon shade. Mark pages through the most obscure sections of the workbook until he finds a question that he thinks will stump John. He likes to play game-show host late into the night and fire off questions at John.

One day, after months of near-silence, John actually lets a remark drop in the kitchen, but, unfortunately, it's not one that scores him points. Larry Gray, a Rescue 2 veteran, has come back to the company as a lieutenant. He was a firefighter under Fred Gallagher and Ray

Downey, and he's got over twenty-five years on the job. Joking around, trying to mimic the verbal jousting of the older guys, John asks Larry, "What is it, Bring Your Grandfather to Work Day?" The audacity of the kid stops the action cold. Larry can't believe this kid is going at it with him.

He laughs and jokes with John, then goes upstairs to plot his revenge. Larry is a talented cartoonist and a witty writer; he dreams up a little pamphlet detailing John's not-too-heroic exploits as chief of the Lakeland, Long Island, Volunteer Fire Department. His drawings show John zooming around in his chief's car just to play with the lights, driving it around town to show off, bossing around guys ten years his senior, and generally making an ass of himself. The guys love it, and the cartoon gets passed around the kitchen. But the fact that they are taking time to make fun of John might also be the first tentative sign that he's becoming a part of the company.

As the new guy, John is assigned the job of can man on every tour he works. This is standard operating procedure at Rescue 2. Giving the new guy the can puts him on center stage. Can carries a fire extinguisher and sticks with the engine company on the hose line. If there's a nozzle to man (which doesn't happen all that often), the can man gets it. Unlike some of the other positions, which operate alone, the can man always goes with the officer and the chauffeur. This way the senior men can watch the new guy and make sure he knows what he's doing. Some guys will tell him right away if he fucks up. Others, like Bob, are a little sneakier. Bob likes to let the new guys think they got away with something. Then, a few hours or maybe a few days later, he'll slowly leak what he knows, watching the new kid suffer each second and hang on his every word.

Several weeks after arriving at Rescue 2, John, still the can man, manages to get the nozzle at a fire in Bushwick after the engine runs out of air. John realizes that this is the moment when he will either prove he deserves his spot in the company, or confirm what the other guys have been saying about him. Dragging the hose, he makes a push around the bend and struggles into the apartment, where the fire is most fierce. Billy Lake, a senior man who is chauffeur this day, watches with satisfaction as he humps hose behind John. After the blaze is out, they stand outside recovering, peeling off layers of soaked clothing, changing out air bottles in preparation for the next fire, and having the usual postgame discussion. Billy comes up to John and pats him on the back. "Nice job, John," he says. Those are his first words to John, and they mean everything.

Two MONTHS after John's arrival at Rescue 2, on a peaceful June night, all hell breaks loose in Brooklyn. Ruvolo, Bob, John, Lincoln, and two other men are working the shift. First, they are called to a gas explosion in Bushwick where several firemen have been injured. The scene, intense and busy, is made more nerve-wracking by the possibility of further explosions. Then, while the Rescue 2 guys are helping to package injured men up for the hospital, Bob hears another job go out over the radio. It's a 10-75 on Atlantic Avenue, right near Rescue 2's quarters, and it sounds like it might be something good.

A few minutes later Bob hears the 10-75 go to a second alarm, signifying that the situation on Atlantic has become much worse. Feeling the adrenaline pump through him, as in the old days, Bob turns the rig around and points it in the direction of East New York, ready to floor the thing and head for the second scene as soon as Ruvolo and the others are on the truck. But before Bob can raise the captain on the handy-talky, the Brooklyn dispatcher calls Rescue 2.

*"Brooklyn to Rescue 2: are you available to take in the third alarm?"* asks Warren Fuchs, the senior Brooklyn dispatcher and a good friend of Bob's. *"We have seriously injured firemen."*

Bob asks Warren to wait ten seconds and tries once more to get Ruvolo on the radio. *"Rescue chauffeur to rescue officer: they've got members down at the second alarm."* Everyone in the company hears Bob's transmission, and the captain leads them through the hectic scene in a sprint back to the rig. Bob gets Warren on the radio and barks out, *"Rescue 2 to Brooklyn. We're on our way."*

As they rush through the still city streets at night, Ruvolo cranks the radio up so as not to miss even a scrap of information about the fire they are speeding toward. It's difficult for Ruvolo to make out the details. Warren is trading cryptic messages about injured firemen with the chief, and it's crazy. Apparently, a collapse has occurred inside a fire in a store, and there are men trapped or severely burned. It's not clear how many, but it seems like a major disaster. From the sound of the radio reports, Ruvolo is going to have to put himself and his men in grave danger tonight. In this kind of situation, the rescue company is a trapped fireman's best bet, the guys he wants coming to help him.

When Rescue 2 arrives, Ruvolo sprints up to Chief Galvin, who is in charge of this fire, and says, "What d'you got?" They stand in front of a three-story apartment building with a large vacant store on the ground floor. Fire shoots out of the second-floor windows and leaps up wildly to the third story. Ruvolo can't see the collapse from the front, so it must be either in the rear of the structure or an internal collapse.

"Phil, we're missing at least three in there," Galvin says.

Ruvolo is concerned. If the building caves in from another collapse, his own men will get trapped. But he makes the call: he has to shake off these doubts and charge in. There are firemen burning up in there. To let his men know just what kind of a job they are facing, he stares into each man's eyes. But Ruvolo speaks calmly to Lincoln Quappe and Billy Esposito: "Lincoln, Espo: you guys go around back and see if you can enter there." Then he turns to John and says, "You come with me." The cool whisper that Ruvolo uses assures the men that he isn't shaken. There's nothing that unnerves a guy more than a screaming officer.

Ruvolo approaches the store, which is dark and smoky and littered with all kinds of junk. The main door hangs lopsided on its frame, already forced open by a truck company. Ruvolo and John step in slowly, testing the floorboards before them, trying to figure out where the collapse begins. After they make their way through one room and pass through an open doorway to the collapse zone, Ruvolo gets a glimpse of the destruction. Somebody's screaming. At least one guy is still alive. Navigating by the sounds of the screams, they move toward the trapped firemen.

The floor is cocked at a forty-five-degree angle, and Ruvolo makes a quick evaluation: The second floor has come down; the fire is raging above. He figures that his men have less than ten minutes in here before the fire chases them out or the collapse buries them too.

Brian Baiker, a probie from 332, is trapped in the rubble. Eric Weiner, a veteran from 111 Truck, is cradling Baiker's head, trying to protect him. Chief Kilduff, a local battalion chief, is trying to figure out how to get Baiker out. Suddenly, Bob materializes out of the smoke, standing over Baiker and trying to pull him out. But Baiker's pinned body won't budge. Ruvolo can feel it: the room is cooking. This rescue has to happen fast or the whole place is going to light up.

Baiker is buried under cabinets, paper rolls, and cardboard boxes. Some of this stuff is burning, some of it is smoldering, and some of it is soaked with water. It's going to take time to excavate Baiker, and there's not much time. The tools are useless, since there isn't room to swing them. In this situation nothing will work but gloved hands.

Baiker lies trapped in a sloped V-shaped collapse. The outer edges of the V stretch up to the ceiling, and the base of the V, where Baiker is, extends to the ground floor. Ruvolo can see the hole above, where the floor gave way. The fire is going so strong over his head that he doesn't even need his flashlight on to operate.

Ruvolo frantically digs. It's very hot, but not smoky here, because the smoke from the fire rises. He knows there's a good chance the floor they're standing on will collapse under them. But he can't do anything about that, except work to get them all out faster. He puts his mind on the pile in front of him and keeps sifting through the debris. John is just to his left, working around Baiker's waist. Bob is to his right, clearing some bigger stuff, boxes and wood, out of the way.

Baiker screams again. "Sir, I know I'm just a probie, sir, but please don't leave me."

Bob almost chuckles as he replies, "Kid, we ain't gonna leave you here, don't worry. We're gonna get you out and bring you back to your wife and kids."

"I don't have any," Baiker yells back. "But you better get me out of here, or I'm never going to have any."

Ruvolo can see that this kid is hurting. The fire around all of them is getting bigger and starting to creep in toward them. But Ruvolo doesn't want to put water on it. The extra weight of the water might make the floor below them give way.

The senior man in the company, Dave Van Vorst, has made a characteristically smart choice: rather than going for glory above, he's climbed down into the basement to check on the beams that support Rescue 2 above him. He radios up to Ruvolo, *The floor beams below you are cracked, but I think they'll hold for a little while.* Ruvolo didn't even think of checking down there, but Dave did it anyway. That's one big reason why a company needs as many senior men as possible.

By now, all the sweat and hot water that Baiker has been sitting in for the past twenty minutes have soaked through his bunker gear. The temperature on the ground spikes up with each ember that falls into the pile around him. He's cooking like a lobster. Baiker can still feel parts of his legs, but he can't feel his feet anymore. He burned his hands trying to put out the embers. As the pain worsens, his nerve endings are flooded with chemicals that his body produces to deaden the pain in his limbs. Soon his whole body will be numb.

From another pile ten feet deeper into the building, Lincoln yells out to Ruvolo that they are making good progress on another trapped man. This second guy is drifting in and out of consciousness. Lincoln can't ID him because he's covered in ash and burns.

John and the captain uncover an air tank as they dig around Baiker. At first they think that it's Baiker's tank, knocked off in the collapse. But then John feels around the front of the tank and realizes that it's another fireman buried facedown in the debris. He isn't moving, screaming, or showing any signs of life.

"No pulse, Cap," says John.

"We take him out first," says Ruvolo, who hopes that they can get in and start CPR on this new victim as soon as they uncover his body. Baiker seems badly hurt but okay. He'll probably survive. But this victim, who is Baiker's officer, Lieutenant Jimmy Blackmore, has to get out immediately, or he's dead. Even if they manage to get him out right now, Ruvolo still doesn't think they'll be able to save him. But it's their job to try.

John is next to Ruvolo, knee-deep in junk, furiously working his gloved hands and spreading the debris away from Blackmore. The heat is increasing. Smoke banks down from above and fills the ground floor. Ruvolo isn't sure how much longer the floor is going to last. Will he see it coming, or will his men end up buried alive, just like this company?

"I should have stayed in the Police Department," yells Baiker. He may be burning up, but he still has a fireman's sense of humor.

Meanwhile an engine company has started spraying water on the fire on the floor above the men. They make some progress at subduing the blaze but, in the process, send burning debris on top of Rescue 2 and the trapped firemen. Ruvolo radios the chief and tells them to turn off the goddamn hoses.

It feels like they've only been in there for seconds, but when Ruvolo checks his watch he realizes they've already been digging for ten minutes. If this was an ordinary collapse and they weren't on top of three dying firemen, Ruvolo would probably take a break outside and reassess the situation, figure out if the fire was going to make the whole place come down. But when firemen are in trouble, there's no such thing as a cost-benefit analysis. That wouldn't be right.

Bob radios for a sawzall tool, a portable electric saw, and keeps digging around Baiker's waist. They'll soon have him free right down to the ankles, and Bob can't figure out why Baiker won't move. Bob tries to yank him out of the pile, but he remains pinned.

When the sawzall is passed to Bob, he revs it up and kneels down on the pile to find what's trapping the men. He starts to cut what feels like a metal pipe wrapped around Baiker's ankles.

"Cut off my legs, I don't care," screams Baiker. "Just get me the fuck out. I need to get out." Ruvolo can't see down to the kid's legs to find out what's wrong. He knows that Baiker is burned up and getting worse as the place heats up. He hopes the sawzall will free him.

Bob leans into the pipe. He thinks it will take a while to saw through the metal. But, with the first incision, the pipe gives way, and a blast of water hits Bob in the face. He feels for the pipe and then realizes that it's only a charged hose line, the one that Baiker was probably pushing into the fire when he fell into the collapse. Now, drained of water, the pressure is released. Baiker's legs come free.

Lieutenant Blackmore, who remains buried facedown, has also been freed, and now Ruvolo and John can move him out. Bob, John, and Ruvolo pass Blackmore, whom they hope, by some miracle, might still be alive, into a Stokes basket, knowing that the paramedics will do everything they can to resuscitate him.

Next comes the fireman that Lincoln and Espo dug out. He's still twisted up in the hose line. Lincoln, unsheathing his knife, crouches down to cut him out. This victim is mumbling incoherently one moment, talking loudly the next. His leg is burned down to the bone.

Last out is Baiker, who screams in pain as they attempt to lift him by his burned hands. He slides right out of his boots, which they leave behind, and is passed onto a backboard. Ruvolo can't tell how badly the kid is hurt. Might be nothing; might be deadly. Burns are unpredictable. Sometimes a fireman will survive horrible burns, only to be killed by an otherwise harmless infection attacking his weakened immune system.

The chief on scene is screaming at Ruvolo. "It's gonna come down," he cries through the smoke. Ruvolo agrees but wishes he could do one more pass of the area to make sure nobody remains trapped.

As the Rescue 2 guys grab whatever gear they can carry and run out the door, Ruvolo breathes a huge sigh of relief that they didn't get dragged into a secondary collapse. He hopes they've saved every one of the men—although it doesn't look good for Lieutenant Blackmore. Ruvolo doesn't expect him to make it, but he still hopes that he might.

Ruvolo emerges onto a sidewalk lined with rigs and uniformed men moving in concert like an ant colony. It's another planet: an open sky above his head, people walking around, calm and able-bodied, not screaming out in pain. Inside that building the minutes flew. The experience passed by in an instant. Now Ruvolo wants to slow it all down and rewind, dissect what they did right or wrong. There'll be time for that tonight in the kitchen and tomorrow over breakfast with the day shift.

In the back of the ambulance, Baiker is screaming in complete agony, "Give me something for the pain," he demands. "I need something for the pain."

"We'll be there soon, don't worry," says the paramedic, who is cutting off Baiker's uniform around the most severe burns. Each piece of fabric peels off a charred layer of skin that sticks to the clothing. They can't give Baiker anything for the pain until they stabilize him at the hospital. They can just try to talk him down.

Baiker sits bolt upright, looks out of the back window, and sees the row frame houses of Brooklyn. The paramedic told him they'd be at the burn center any minute, but they're still in Brooklyn. "Liar!" he screams at the paramedic.

Down on the street one of the engine guys tells Ruvolo that two more injured firemen had been dispatched to the burn center before Rescue 2 arrived. That means five firemen badly injured, maybe dead.

Gradually, as the Rescue 2 guys hear what happened before the collapse, they piece the whole situation together. It all started with a report of someone trapped in an apartment above the store. "My mother's in there," an African-American man had yelled as he ran down the sidewalk in front of the building. "She's on the second floor."

Immediately, the chief had ordered Baiker's company up to the second floor to knock down the fire so the truckies could find the woman, who later turned out to have been outside all the time. After they pushed up there, the floor cracked with a sound like a gunshot, and everything gave way underneath. Seconds later, Baiker and the rest of his company were buried just where Rescue 2 found them.

When Bob starts talking to the men from other companies, a fireman tells him that Timmy Stackpole was the guy Lincoln and Espo pulled out. Bob can't believe it. He didn't even recognize Timmy, a former Rescue 2 fireman, now a lieutenant in 103.

As Ruvolo leans on the back of the truck and takes off his soaked gear, he sees Ray Downey pull up in his chief's car. Downey comes over, and, without preliminaries, Ruvolo tells him, "I think we lost at least one, Ray. We pulled two others out who looked okay." Downey nods, and then heads over to talk to the other chiefs.

On the way over, Downey sees the men from 176 Truck huddled together behind their rig. He knows their officer is seriously injured and probably going to die. He doesn't know who that is.

"What's the name of your captain?" he asks them.

"He's a covering guy."

"It isn't LaPiedra, is it?"

"That's him."

Downey had chosen LaPiedra as the newest officer in his Special Operations Command. Scotty LaPiedra, who was sent to the hospital before Rescue 2 arrived, was working his last tour in a truck company tonight. He was due at his new job tomorrow morning. He was a guy Ray Downey had seen at work as a fireman in Squad 1 and hand-picked as a future rescue officer.

On the drive back, Bob and Ruvolo are silent. Ruvolo is going over the what ifs—*is there any way we could have gotten Jimmy Blackmore out quicker or started CPR on him?* Ruvolo has pulled firemen out before, but he has never been in something like this. On top of the difficulty of getting the guys free, they had a fire raging above them. Ruvolo can't believe they didn't get caught in another collapse. Who would have come for them?

At the firehouse that night the men are quiet. Bob makes up his bed and settles down. He can't fall asleep, but he also doesn't want to stay awake. He just wants to lie there in the dark, not dreaming or thinking of anything. The Atlantic Avenue Fire is the kind of job that Rescue 2 is made for. It's the kind of job that comes along once in a lifetime. It was a rescue of brother firemen under the most intense pressure imaginable: a fire raging around them and the constant threat of a secondary collapse that would throw victims and rescuers into the inferno. Everyone in the FDNY will know about it soon, from the chiefs in headquarters to all the guys in the local companies. Previous captains, like Fred Gallagher and Ray Downey, had proved themselves by rescuing firemen, just as Ruvolo did at Atlantic Avenue. Maybe this is the beginning of an equally legendary career for Ruvolo. But after a day like today you don't give a shit about legends. You just remember the guys screaming, and the ones who died.

A few days later there is the funeral for Jimmy Blackmore and visits to the hospital to see the injured firemen. When Bob and the captain go into the burn center, Brian Baiker yells out in surprise when he sees Bob. He didn't even know the name of the guy who helped save his life. All he remembered was Bob's distinctive mug, seen from below. Bob tells him, "Hang in there, kid. You'll be all right."

Timmy Stackpole, whom Lincoln and Espo dug out, is wolfing down fruit by the carton when Bob and Ruvolo arrive at his room. Already he's talking about coming back on the job. Bad burns accelerate the metabolism so much that, if they're not careful, Baiker and Stackpole could easily lose fifty pounds each. Baiker has already lost twelve pounds in just one day. When Timmy talks about coming back on the job, Bob looks down at Timmy's feet, burned to the bone, and thinks it will be tough to pull on a pair of fire boots ever again. He wishes Timmy the best, but he wonders if maybe his friend should think about retiring. After all, he's got a wife and three kids worrying about whether he'll live through this.

Down the hallway Scotty LaPiedra, the captain whom Ray Downey had picked for his rescues, is hanging on by a thread. By a miracle of modern technology the doctors have managed to give him a few extra days on this earth. Those days are a chance for his family to say good-bye, but the man they are saying good-bye to is burned badly, almost beyond recognition. Kevin O'Rourke was one of LaPiedra's friends at Squad 1 in Park Slope, so he and his wife, Maryann, make a pilgrimage to the burn center. After they walk out of there, tears stream down Kevin's face. "God forbid that ever happens to me," Kevin says to his wife. "I don't ever want my kids to see me like that." That's all that he says to Maryann, but it's enough to make her worry about him. Strangely, it's not the thought of his getting hurt that preoccupies her; it's the thought of him going to work thinking of Scotty LaPiedra.

JOHN GAINS A NEW RESPECT in the firehouse after the Atlantic Avenue fire. His days of being the quiet kid that no one spoke to are almost over. Ruvolo is proud of John. He wasn't sure about him at first, and he wondered if the kid would survive the intense hazing. But now he calls John into his office. John enters sheepishly, afraid of another speech like the one he endured when he entered the company. But things have changed. This time Ruvolo simply tells him, "John, I'm glad you're here."

In fact, Ruvolo is starting to like John, whose quiet determination he is beginning to admire. Ruvolo was predisposed to hate this kid. He was ready to see him as an undeserving, spoiled prick. But John has proven Ruvolo completely wrong. When he thinks of John the same phrase keeps coming to mind: "excellence, understated." And the kid has a warmth about him. Although he hasn't said much so far, he can deliver a pretty good line. Now, looking at John, Ruvolo starts to see a little bit of himself in this younger firefighter. John is serious about studying, just as Ruvolo was. The captain has noticed how hard he works. He even looks a little like Ruvolo did, ten years and twenty pounds ago.

When John takes the lieutenant's test, he passes with flying colors, scoring in the 90s, higher than Ruvolo scored ten years ago. But John has so little seniority that he might not get promoted. He'll have to stick around at Rescue 2 and wait for the next test, three or four years from now. That's okay with John. He's starting to love this place. Ruvolo thinks that the time will be good for John, a chance for him to mature before he has to lead other men.

Meanwhile at the kitchen table Lincoln Quappe plunks down his copies of the books to study for the lieutenant's exam. He even takes a look at them occasionally. But he has a different sort of ambition from John's. Lincoln just loves being a fireman. He doesn't want to be a boss. Sure, he'd like the extra few grand a year, but he likes his current job too much to angle for a promotion. The old days in his previous firehouse when he used to complain about the Rescue 2 guys seem long past.

At Mark Gregory's birthday party, Lincoln has to face the wrath of his wife, Jane, when he hints that he might want a Rescue 2 tattoo like many of the other guys. "You're not getting a tattoo," she tells him, right in front of Richie and the rest of the men.

"What are you doing to me?" he asks her later, on their way home. "You can't say that in front of the guys."

"You're *not* getting a tattoo."

Lincoln has also gotten tight with Mark Gregory. He loves to get Mark in trouble and then chuckle about it on the drive home. In the middle of a tough fire, when they're on their bellies crawling down a hallway, Lincoln will start talking about carpooling, just to crack Mark up. They often commute in together, and Lincoln always brings a minicooler of beer to drink on the way home. Sometimes he'll buy the weirdest, cheapest beer he can find and hand Mark a can of that while he sips his Budweiser. He puts out his cigar right before he gets home, and he always brushes his teeth before kissing his wife. He also makes a point of stopping at a trash can near his house and tossing the empties. But she can still taste the cigars, and she does find the occasional beer can in the backseat and chastises him for it.

Lincoln doesn't tell his wife anything about the Atlantic Avenue Fire. He must figure it would upset her too much. As with many of the other Rescue 2 firemen, for Lincoln the division between work and home is absolute. At home, he is Mister Mom. He even arranges his shifts so that his wife can work while he takes care of their two kids, Clint and Natalie, on every free day he has. But at work he cultivates a totally different personality, a kind of Long Island redneck, a squirrel-hunting character from the film *Deliverance*. The guys who get to know him, like Mark, realize that's just a good act and that he's laughing all the way, while some other guys mistake him for a hick or a clown.

That same June, the off-duty men of the company travel to city hall for the annual Medal Day ceremony. The park next to city hall is reserved for the firemen and their families. Thousands of chairs fan out from the steps. On these occasions, the department and the city hand out gold medals and cash prizes to the firemen who have made impressive rescues in the last year. The Rescue 2 guys always have someone getting a medal, so it's become a company tradition to pose on the steps with the bulldog banner and snap a company photo. This year, Bob Galione is receiving the Brooklyn Citizens' Medal and the recently instituted Louis Valentino Award.

On this sunny June Medal Day, Bob thinks of two things: Louis and his rescue of the old man in last year's building collapse, which is why he's getting the medal. But Bob doesn't want to think too hard about Louis today. He's got to keep it together when he goes up to get his medal.

So he concentrates on the many old friends whom he sees at the ceremony, men from Rescue 2 and other companies, men whose faces he has encountered at fires for as long as he has been a fireman. He also thinks of the old man he saved that day and what that rescue did for him in the dark days after Louis' death. As much as Bob likes to say that his job is about adrenaline, fear, and fun, there are also many days like today, when he loves this job because it gives him the opportunity to help others, especially old people, kids, and the disabled—the ones who can't get out by themselves. And as much as he doesn't want to admit it, part of what makes him get up every morning for work is the chance to save a life. He's grateful for being given that chance.

At around the midpoint of the ceremony, Bob steps up to the podium to shake Mayor Rudy Giuliani's hand and receive his medal. The trip up to the stage feels like a very long road to him. Then suddenly Mrs. Valentino is draping the medal around his neck. Before she can speak, Bob whispers in her ear, "Don't make me cry, Mrs. V.," he says. "Please don't make me cry, Mrs. V."

# FIVE:

## THE EIGHTIES

**P**HIL RUVOLO HAS ESTABLISHED a morning ritual: every day at 7 AM, from his upstairs office in the firehouse, he picks up the phone and calls Ray Downey to talk shop, shoot the breeze, piss and moan, and take stock of the world. In the seventies, Rescue 2 belonged to Fred Gallagher, but from 1980 to 1994, it was Downey's baby. In fact, Downey selected most of the guys who serve under Ruvolo. When Ruvolo says, "The boys made some enemies last night," Downey knows what he means and he can visualize just who shot their mouth off or who got a little physical in the fire.

Downey is Ruvolo's boss and the chief in charge of all five rescue companies, seven squads, the Hazardous Materials Unit (Haz-Mat), and the Marine Division (the fireboats). Downey takes Ruvolo's calls in an administrative office surrounded by filing cabinets, computers, and office supplies. He only gets out in the field for major disasters, so these phone calls connect him to the reality of life on the job and the daily routine of his old firehouse. Hearing Ruvolo's voice in the morning, being able to picture him sitting at the same wooden desk where Downey himself used to sit, means something to Downey. He's a man who values order and continuity. And he'll never forget his days at Rescue 2.

Ray Downey is an Irish-American from Woodside, Queens, a neighborhood once filled with rows of almost identical Irish bars where the regulars started drinking at breakfast and didn't stop for lunch. These bars were poor men's country clubs, where a guy could find work, make connections, or get into a brawl. Little kids stepped up to the bar carrying metal pails called growlers, which the ruddy-cheeked bartenders would fill up with draft beer for fathers drinking at home. Forty-eighth Avenue, the main thoroughfare near Ray's house, had three funeral homes and a dozen bars. Large and small apartment buildings filled the nearby streets, along with some two-family homes. Hardly any families had a full house to themselves and no one could escape from the shadow and squeal of the Number 7 train, which slowly snaked its way to Shea Stadium and finally on to Flushing. In the old days the windows of the elevated IRT revealed nothing but white working-class people, far down the line, but now these neighborhoods have been transformed. Jackson Heights, just past Woodside, has countless storefronts that sell empanadas and tacos. Flushing, the end of the line, has so many Koreans and Taiwanese that old-timers call the Number 7 the "Orient Express" or the "Slow Boat to China." The old white, working-class neighborhoods, places like Woodside, Flatbush, Bushwick, and Kingsbridge, have been largely transformed into neighborhoods for the city's newest immigrants.

Ray Downey was born in Woodside in 1940 to working-class people who had slogged their way through the Great Depression and struggled through World War II. Their own lives had made them even more determined to make something of their four sons, and their one daughter, Alice. (Ray was the youngest.) Ray's father, Joseph, was a lift operator who ran elevators at construction sites, but during the thirties there was little construction work in the city. The best he could do was to join Roosevelt's brainchild, the WPA, and travel out to New Jersey, doing whatever he could to bring home a few bucks. The family moved from apartment to apartment, always on the lookout for a better deal, and the boys shared one room, escaping from the dormitory only by getting married.

In Ray Downey's Queens, as in Fred Gallagher's Brooklyn, neighborhood was defined by parish. Downey went to St. Theresa's, a

Catholic church where the nuns had names like Sister John William and knew how to use a ruler or throw a punch. In the first grade, Ray and his friend Artie Weisenseel were marching outside with their classmates. When the nun told each boy to pick a girl to hold hands with, and get in line, Ray and Artie chose the same little girl and wrestled each other to the ground, fighting over her. Sister Rose Delima broke up the scuffle and said, with a thick brogue, "Boys, stop. I have the solution." She called another little girl forward, the identical twin of the girl they had been fighting over. Both boys got their girl.

Sunnyside, within walking distance of Woodside, was a bedroom community where it was rumored that rich men housed their showgirl mistresses. Ray's older brothers cautioned him about some of the bars over on Queens Boulevard whose bookies, hookers, low-level criminals, and hoods hung out alongside honest workingmen. Artie's Jewish uncle stowed betting slips under Artie's bed—the cops wouldn't search a little kid's bed.

Woodside was a world of fierce loyalties—to the Church, to the country, to Ireland, and to the honor of Woodside itself. When Ray was little, the Woodside postal code was rumored to have the most Congressional Medal of Honor winners in the United States. And the whole neighborhood turned out to welcome Frank Smith home from World War II, where he had led the charge up Sugarloaf Hill. Every boy should serve the country, people in the neighborhood believed, and that meant time in the military. After they returned home, most Woodside men took up a trade, often one that ran in their family— like steam fitting or lathing. Each union had a reputation. The lathers were rumored to pack weapons and be ruled by ex-cons. The transit workers had lots of IRA men on the payroll, guys wanted by the Brits, hiding out in Irish America. At school, along with the "Star Spangled Banner," Ray learned the Irish Anthem.

A Woodside childhood was filled with rough games and crude comforts. Roller hockey ruled Woodside. There were five leagues, from the Rebels and the Junior Americans on up to the Americans (rumored to be paid to play by local bookies). The equipment was simple. A roll of black electrical tape, greased to move, served as the puck. Ray and Artie started out with clip-on wheels but lobbied their parents for shoe skates. The kids raised money for team jerseys by hawking raffle tickets in the neighborhood, a nickel apiece.

Ray was a standout on the team; roller hockey was his life. He and his friends on the neighborhood team trekked across the city to play, traveling as far away as Hell's Kitchen and Red Hook. After their team won the city CYO and YMCA championships, Ray and Artie felt certain they would be professional roller-hockey players in the future, though back then even big-time athletes didn't make much money. Many members of the professional ice hockey team, the New York Rangers, lived right in the neighborhood, in humble homes.

Ray's stop on the Number 7 train was the most elegantly named station in the city—Forty-sixth Street Bliss. (Now it's plain Forty-sixth Street, no Bliss.) Around the corner from the stop, down on Queens Boulevard, was the Knights of Columbus, known as the K of C, a Catholic fraternal organization that hosted community dinners and served as a meeting place for locals, who rarely left the neighborhood. Farther down Queens Boulevard, set back from the street, was an exotic animal warehouse that enchanted neighborhood kids, who were always trying to peek in and see a giraffe or an elephant. There were big brick housing projects, dubbed "the Mets," because they were built to house Metropolitan Life's workers. There were endless rows of identical two-story houses with no yards in between, only a back alley running behind them. In these alleys Artie and Ray would roast potatoes over a cooking fire they built. The potatoes were called Mickeys, because it was the Irish who ate them, and they were best served burned with a big piece of butter shoved inside.

There was no Halloween holiday for kids in those days, but on Thanksgiving, Ray and the others would blacken their faces with ash and dress up like bums. "A penny for Thanksgiving," they'd yell in the streets, and some apartment dwellers would toss out a couple of coins. A few cruel adults might heat up the pennies so the kids would get a jolt when they picked them up.

Young boys in Woodside congregated first on the street corners and later at the bars, where fistfights were common. Ray watched one of the most famous fights in the neighborhood, in which young Buddy Bear faced off against an older, barrel-chested barkeep. The two combatants downed shots of whiskey between rounds, and, by the end, Buddy had punched himself out of commission—like George Foreman when he lost to Muhammad Ali. Ray became a leader and a formidable fighter, tough even by Woodside standards. But while it was okay to face off against kids in the neighborhood, or even the rowdy Italian kids from other Queens neighborhoods like Astoria or Corona, when Ray's gang headed to Red Hook, Brooklyn, to buy fireworks, they backed off a little. The kids there looked *really* tough.

Afternoons were filled with sports—roller hockey in quiet streets next to the cemetery, football in the potter's field, the big open area inside the cemetery where the anonymous poor were laid to rest. Ray would also go to the graveyard on Memorial Day and Veteran's Day with a small hand shovel and a can of water. He'd help people plant flowers next to the graves of their loved ones and make a few bits in tips. He went on to high school in Manhattan at La Salle Academy, a Catholic boys' school on the Lower East Side. But the pull of Woodside stayed strong. In the afternoons he played sports in the old neighborhood: eventually he moved from roller hockey to football and quarterbacked for the Rambucks, who became Queens champs. In his free time, he chased girls in Woodside and in Rockaway, also known as the Irish Riviera, where Italians infringing on the Celtic turf knew they'd face a fierce fight.

RAY DOWNEY and his family always lived in the same few blocks of this comfortable, old, conservative neighborhood, a place where it wasn't uncommon to see an Irish Catholic kid decide to make his living fighting fires. By the time Ray was twelve, two of his three older brothers had joined the Fire Department, and he fully expected that the day was coming when he would take his place there too. He was more or less brought up for the life: The Downeys were a straitlaced family. Ray's father, Joseph, had always dressed for church and never drank to excess. Joseph Downey had died when Ray was just nine years old. The funeral was at Hughes' Funeral Home, one of three parlors just down the block. All the families in the neighborhood crowded in to pay their respects, and, later, everyone retired home to mourn their dear departed. Ray's older brothers had lived at home as he was growing up, shadowing his mother as older sons did in those days and helping to make up for a father's lost wages.

Just as patriotism, fidelity, religion, and combativeness seemed to be in the water, two opposing traits also marked most Woodside families. There were garrulous clans of storytellers and drunks, showboats and showmen, who thrived on talk. Then there were clans like the Downeys, who were as tight-lipped as could be. Along with everything else he got from his older brothers and his parents, Ray got this trait.

At nineteen, Ray shipped out with the marines to see the world before he followed his brothers onto the job. It was the peaceful time between Korea and Vietnam. Ray was stationed in Cyprus. It was to be the only extended period of time he would spend outside the city he loved. When he came back to Woodside, the neighborhood hadn't changed much. Shortly after his return, he did battle outside a bar with Tom Burns, another Woodside kid, and dislocated Tom's shoulder in the fight. Later Ray faced down an Italian kid who would go on to lead the Arizona mafia as an adult.

When Ray came back from the marines, he was ready to join the FDNY. After passing the test, he waited for his name to reach the top of the long list. In the meantime, he found a boring job at a New York Trust Company office. The job did have one advantage, though: if he looked up from his desk, through a glass partition, across a hallway, and through another glass partition, Ray could catch a glimpse of Rosalie Princiotta. An Italian-American girl from Brooklyn, she had long black hair and a smile that could make Ray lose track of his work. To an Irish-American guy from Woodside, Rosalie was as exotic as a foreign movie, but Ray passed her a note anyway, and they went out on a double date with another coworker and his girlfriend. Their first date alone came later, on a Friday night at a bar in Woodside called the Log Cabin. It was Lent, so Rosalie told her mother she was going to an evening mass after work; her mother didn't approve of Rosalie going to bars.

They started dating in the winter and Rosalie could tell Ray was serious because he was willing to trek over to Bushwick every weekend to see her. The hour and a half trip required two buses and a subway. But the relationship screeched to a halt when Rosalie saw an ex-girlfriend's name tattooed on Ray's bicep. She hadn't noticed it in the winter, but when spring came and Ray put on a T-shirt, Rosalie was horrified by that other girl's name on his arm. The next day he came into work and unwrapped a bandage to show Rosalie what he had done: the other girl's name was now tattooed over with roses.

Back in the sixties, Italian-Americans and Irish-Americans lived in separate worlds, kept apart by strong mutual antagonism. The Irish were always griping about how the Italians would work for cheap and take away their jobs. The Italians saw the Irish as lazy drunks and fighters. In truth, the communities had much in common—religion, a strong work ethic, patriotism—but if you implied as much to an Irish or Italian guy, you'd probably get a punch in the nose for your trouble.

Rosalie lived in an Italian neighborhood in Bushwick, in a so-called coldwater flat. When someone in her family wanted to take a bath, they put a pot of water on the stove and warmed it up, then bathed in an old cast-iron tub anchored in the creaky old kitchen. Rosalie's family, especially her father, took to Ray immediately, even though he definitely wasn't Italian. When the holidays came around, Ray worked in her father's grocery store in Williamsburg, selling Christmas trees. Many of the old patrons didn't speak English, so Rose's dad told Ray, "Just say 'venti cinque' when they ask," which was twenty-five in their southern dialect.

After Rosalie, Ray never looked at another girl again. He was a decisive guy—and he had decided. Within two years he and Rosalie were married. Her family had a summerhouse in Deer Park, in Suffolk County, Long Island, and when Rosalie's Brooklyn neighborhood started to go bad, her parents moved out to Suffolk full-time. They built a brand-new house in Deer Park, a community being built from the ground up that seemed fresh and beautiful to a couple of city kids like Ro and Ray. The selling price was $13,999. Rosalie's parents picked out their plot of land and put down $50 on the house, which was a split-level big enough for two families. Ray and Ro could live downstairs, and her parents would have the upstairs.

Gradually, Rosalie's extended family would follow her to Deer Park. For Ray and Rosalie their new block became an extension, almost a replica, of her old neighborhood. Back in Bushwick six families had lived all together in one big apartment building. Five of those families now moved to the same block in Deer Park. Having Rosalie's family right there on Oak Street meant that when Ray Downey had to work nights, he knew that his wife was never alone. She was protected on all sides by people they knew.

There were no small parties, families, or appetites on Oak Street. If somebody had a birthday, a communion, or a graduation, there was a whole block full of relatives to invite. Ray blended easily into the new community and became an honorary Italian-American, as comfortable drinking a shot of grappa as a glass of whiskey. When they threw parties, Ray and the men pitched giant tents out back, and the women brought out huge heaping foil trays of southern Italian pasta and gravy. On these nights, the old Italian men would gather around the dining-room table upstairs, drinking espresso or chewing on fennel. The men sat all together at one end of the table, playing cards, and the women sat at the other end, drinking coffee and eating cake. To these old men Ray was a good boy, a hardworking fireman. He didn't play cards, but he sat with the men out of respect.

Ray and Rosalie's experience of a big family so nearby was unusual for the suburbs; for most people, moving from the city to the suburbs meant losing the intense communal life they had before. Even after Ray Downey left Woodside forever, the ethos of the neighborhood stayed with him for life. Woodside boys led the charge. They didn't back away from a conflict. They didn't trust outsiders. They believed that some things were worth not just fighting but dying for. The values of Woodside were, in essence, the values of the FDNY. They were also the same values of the many white working-class neighborhoods that dotted the outer boroughs, all those places where Ray went to play roller hockey.

Over time these neighborhoods would shrink and mostly disappear. Now only a few old-timers still sit on the barstools of Woodside and most working-class white people live in the suburbs. Few of the old neighborhoods live on today. But the spirit of these places still defines the FDNY. To men who grew up in these neighborhoods, the job feels like home. Even firemen who grew up in the suburbs are often the sons of men and women from these urban places. For them, belonging to the FDNY is a little like going back to their fathers' old neighborhood. In the FDNY, as in the Woodside of Ray Downey's childhood, making money isn't considered much of a goal, and serving your country is almost a requirement. The old white working-class spirit of Woodside may not have a geographical home anymore, but it lives on in the FDNY.

RAY DOWNEY reported for his first day of work as a fireman on April 7, 1962. His destination was the fire academy on Welfare Island, a narrow strip of land just east of Manhattan dotted with abandoned buildings and a mental asylum. Like his predecessors, Downey slaved away cleaning up equipment, trained in basic procedures, got sent into smoke-filled rooms, and quickly picked up the lingo of the job. Eventually, he and his pals were sent to Manhattan to finish off their training, where they roped down six-story buildings, in preparation for work in the busy metropolis. After eight weeks of training, Ray and his buddies were sent off to their first firehouses.

In the fall of 1962, Ray found himself at Engine 35, uptown in Manhattan. He knew right away that he wanted a firehouse with more action. His goal was to get himself to Ladder 4, in the busy heart of Midtown. After cutting his teeth at 35 for a year and a half, he finally saw his ticket out: an angry old battalion chief needed an aide to drive him around. Normally, this was a job for a senior man with ten or fifteen years in the FDNY, but none of the experienced firemen wanted to drive this particular chief, who was a curmudgeonly pain-in-the-ass.

So Downey got the job. He did as he was told, without a complaint and a year and a half later, with that chief's recommendation, he was assigned to Ladder 4.

And so Downey became a real working fireman, racing to old tenements that lined the blocks west of Times Square with another gung ho junior guy named John O'Rourke, who would later become chief of department. Times Square then was a far cry from the Disneyfied version of today. It was peopled by junkies, hookers, freaks, and soap-box preachers in addition to the city kids looking for a good time. Back then, all the first-run theaters were right in Times Square. During those years, the place pulsed with an energy that was unique in the world—highbrow, lowbrow, and no brow all meeting at one corner, twenty-four hours a day.

On one memorable occasion, Downey and O'Rourke caught a subway fire burning intensely down below street level. When the two crawled down to check it out, they saw a carton of dynamite near the blaze, escaped quickly, and reported it to the chief. "Go down and get it," the chief barked. Too naive to protest, the two young firemen crawled back into what was by then a large conflagration and dragged out the dynamite. They got out without an explosion.

After one shift at Ladder 4, Ray came home and told Rosalie about his day. He had hung off a ledge by his fingers, almost plunging to his death. "Ro, it was the first time I saw my whole life before me," he told her. "I really didn't think I was gonna see you or the kids again." Rosalie got so upset that Ray never told her a fire story again— for thirty-five years. But he couldn't keep her out of the loop altogether, since periodically, Rosalie would get a call from the firehouse asking her to come and pick Ray up at the hospital. She would bundle up all the kids into her brother's car and drive into the city. Once, she waited for half an hour to enter the ward to see Ray. The doctors took her to a bed, and she stared at a blackened body. She shrieked, then looked at the man's face. And then she realized: It wasn't a blackened body; it was a black man. They'd taken her to the wrong bed. Ray was down the hall, as white as ever.

One New Year's Eve a car pulled up in front of the Downeys'

home in Deer Park. Two guys got out, escorting Ray. Watching from the front door, Rosalie thought he must be drunk. But when he got to the door, she realized that the smoke had temporarily blinded him. They sat Ray down at the head of the table and the rest of the family gleefully played tricks on him all night. They handed him a plate of food, then secretly moved it. "You guys are one sick family," a relative remarked to the Downeys. But it was just their way of having fun with the situation.

Like most firemen, Downey worked a full array of side jobs to make ends meet. Rosalie teased him, calling him "Jack of all trades, master of none." He drove school buses, painted houses, worked as a lather, a carpenter, and a department store clerk. On top of all that, he had a full course load at Suffolk County Community College. Other firemen might study part-time, but not Ray Downey. He could only play hockey for the school team if he was a full-time student, so that was that.

After a few more years as a fireman, Downey—who had studied hard for promotion during every free moment—passed the required examination and made lieutenant in 1969. He was a fireman on the rise, but to anyone who looked closely it was clear he was still part marine. You could spot him a block away by his ramrod-straight posture, which had not slackened since his days at boot camp on Paris Island. He had a habit of looking his fellow firemen straight in the eye, and though he spoke little, he said exactly what was on his mind. All this made him a remarkably effective commander, though he wasn't the kind of officer that firemen necessarily *liked*. He was too tough for the guys to like him. And he didn't care whether they did. From the beginning he had an intensity that could induce fear. He demanded a lot of his men, and he got it. His manner might be intimidating, but the men respected the attitude behind it: Ray Downey never seemed to forget just how serious this job was.

After seven years as a lieutenant, Downey made captain and got the plum assignment of opening Squad 1 in Park Slope, Brooklyn. Squad 1 was to operate out of the old quarters of Engine 269, right in the heart of Park Slope. The city had tried to close down 269 to save money in a fiscal crisis, but politically connected Park Slope citizens had lobbied their friend Governor Hugh Carey, persuading him to reopen the unit. The city authorities couldn't just reopen the engine company; that would be admitting defeat at the hands of the governor. So to save face they called the company a "squad," and designated it an additional-manpower unit to help in covering busy areas. In the late sixties, there had been nine squads all over the city that rotated around to supplement the busiest firehouses with fresh men, giving much-needed relief. But by 1977, they had all closed down. Ray Downey was handpicked to lead the first of the new squad companies.

Squad 1 became Downey's own unit, built from scratch, something he could take pride in. He picked guys he liked and who liked him. They rode with some of the same tools as the rescue units but also packed engine equipment so they could respond as a regular first-due engine company. When Downey opened up shop in 1977, Squad 1 got a brand-new rig, the fastest around. The squad even beat Rescue 2 to some jobs because their truck was so fast. An intense rivalry developed between the established, elite unit, Rescue 2, and Downey's squad of young upstarts. The competition was fierce. At one fire in Flatbush, Jack Pritchard tried to run Downey over with the Rescue rig when he saw the squad guys beating them to the fire. Jack honked the horn, Downey glared, and then Jack put his foot on the gas, just for kicks. He knew they'd get a scare and move out of the way. Downey dived into a snowbank just in time to escape injury: the two men hated each other after that.

Downey delighted in the practical creation of his new company. Following a long Fire Department history of scraping and scavenging for everything, he managed to fully outfit and equip the company in a matter of months. On opening day, when one of the new guys brought in an enormous sheet cake, Squad 1 didn't even have a knife. They had to borrow one from a neighbor to cut it up. But within a few months, it looked like they'd been in the firehouse for a decade.

Squad 1 chose the phoenix as its emblem, and out of the ashes of the past, its firefighters built a highly respected company. Men came from all over the city to join Squad 1. Some were eager to work hard, as Downey demanded, but others were drawn by the old squad companies' reputation as havens for guys who liked to drink and party at work. That didn't wash with the new captain; he didn't tolerate guys who fucked around. Within a year he had weeded out the undesirables and was in command of the company he wanted.

The men of Squad 1 found these early days exhausting and high pressure, but also idyllic. For Al Washington, one of a small number of African-American firemen on the job in the seventies, it was a welcome relief to be in a company of men who treated one another as equals, a value that Ray Downey emphasized. Al hadn't always found himself among fair-minded men: during his time at the academy, black recruits were routinely treated with such viciousness that one of Al's African-American friends took to carrying a .38 inside his fire coat, just in case he had any trouble. Al had been threatened and bullied so often at the academy that he warned his wife not to believe what the department told her if he got killed on duty. He wouldn't have been surprised if somebody had tried to push him off a roof. At his first firehouse, there had been guys who hated his guts. On the first day he showed up for work, there was a "No Niggers and Spics Wanted" sign on the blackboard. And yet, in a fire, every one of the men in his first company would have risked his life to save Al's life. But he knew he'd never be their friend. At Squad 1 things were different. These were men who might become his friends.

Though they worked hard to build up the company, the men of Squad 1 remained typical Brooklyn boys and were not immune to life's pleasures. Every payday, which fell on Thursdays, fifteen or twenty Squad 1 firemen would go out together in Park Slope. Within just a block or two there were half a dozen Irish bars and a couple of black bars. Their first stop was always Mooney's on the Seventh Avenue strip, where the proprietor was willing to cash all Fire Department paychecks. He'd take all the checks in his hand and then go downstairs to get the money, returning with a huge stack of bills, which he counted out to the firemen waiting to drink up a little bit of their earnings.

AFTER HIS SUCCESS with Squad 1, it was only logical that in 1980, when Fred Gallagher made chief, Ray Downey would get the captain's spot at Rescue 2. Whatever problems he had encountered at Squad 1, Downey knew right away that this new gig was going to be something else completely: Fred Gallagher's tribe made just about any other firehouse in New York seem tranquil. Gallagher had left behind a group of men who still worshiped him, guys who weren't ready to welcome a new captain, especially not that bossy SOB from Squad 1.

When Gallagher left the company, some of the best rescue firefighters felt like they had lost a father. He had taught them how to fight fires. And they had had tremendous fun doing it. They loved the guy. Gallagher might demand top performance at a fire, but he let the men do whatever they wanted in the firehouse. He joined in food fights and wrestling matches, where he had as much fun as his men, while still remaining their captain.

Downey was a very different kind of captain. He wasn't going to be giving out bear hugs or tackling men in the kitchen. He wasn't much of a talker; in fact, he didn't say much of anything except when he was barking out orders. But even if Gallagher's boys could get past all that, there was one thing they couldn't change: Downey wasn't Fred Gallagher, and therefore they would never accept him as captain. When Downey had to stay overnight in Brooklyn, but wasn't working, he'd drive down Eastern Parkway to Squad 1 and bunk there. For a long time, that still felt like where he belonged.

It wasn't long before some of Gallagher's men at Rescue 2 were making plans to ship out. Jack Pritchard got promoted and left Rescue 2. Tom Dillon retired with twenty years behind him. Slowly, Downey started to bring in his own men to Rescue 2, men like Al Washington, John Barbagallo, and a few other guys he had worked with before.

Al Washington was reluctant to follow Downey to Rescue 2, where Bill Floyd, another black firefighter, was retiring. Al was happy at Squad 1 but if he didn't take Floyd's place, there would be no black firemen in any of the rescue companies—a fact that troubled the

THE LAST MEN OUT

Vulcan Society, an organization of black firefighters. These men urged Al to go to Rescue 2. At the urging of his peers he agreed to go into Rescue 2, but not without some trepidation.

It was a decision he didn't regret. Months later, Al was working a job in Bed-Stuy where a woman was trapped in the flames. She was on the phone with the dispatcher but couldn't get out. Leaping over a fence, Al sprinted around to the back entrance of the fire building and charged upstairs to search the bedroom. There he found the woman passed out and dragged her back through the hallway to safety, then went back in to fight the blaze. By the time he was finished, she had been whisked away to the hospital. A few months later, Al got a thank-you note, and then ran into the woman he had rescued near the fire-house. After Al got off work, they took a bike ride to Brooklyn Heights and had ice cream cones. Eventually, they became best friends, and Al moved into the apartment where he had saved her life.

Another of Downey's stalwarts was John Barbagallo, a legendary old-school nozzle man who had worked in East New York's Engine 290 in the sixties and seventies before transferring to Harlem. John first met Ray at 58 Engine, a legendary Harlem company dubbed "The Fire Factory." Ray Downey admired the fact that even after twenty years on the job, John still had all the enthusiasm of a probie. When he came out of a fire covered in ash, all burned up, the first words out of his mouth were always, "This is the best fucking job on earth." Then he'd crack a huge smile, sit back on the rig's front bumper, and relax. When he ran into new kids outside a job, he'd excitedly approach them. "Hey, kid, how 'bout I give you my twenty years on the job, and you let me start all over again?" Sometimes he would wink conspiratorially and say, "Who told you about this job?"

At Rescue 2, John wowed the guys in the kitchen with his stories of the war years in East New York. But some wise guy always piped up, "Hey, John, if you guys did such a great job, how come the whole place burned down to nothing?" John didn't care what they said. He wasn't there for them; he had come to Rescue 2 because he considered Ray Downey to be the finest fire officer he had ever worked with.

Right off the bat, Downey made John his chauffeur, a decision that almost backfired. Though Downey normally was not a patient man, he had seemingly infinite patience with John's less-than-precise navigating. John knew Brooklyn as a firefighter. Burned into his brain was the blueprint of every kind of building you could come up against in the borough. But outside of a fire building he was lucky if he could find his way across the street, a liability that Downey hadn't known of when he gave John the chauffeur's seat.

Luckily, however, the captain's new chauffeur had a few tricks up his sleeve: If John saw a TV news van speeding alongside the rescue truck, he would follow it to the fire. Other times he would crane his neck and look for the smoke. That worked fine until they got close to the scene; then the buildings would obscure the plumes of smoke, and he'd be officially lost. The guys in the back could tell they were getting closer because their fire ground radios would start to buzz with on-scene communications, but then, when John made a wrong turn, they would lose reception. They'd try and yell out directions, but up in the closed cab, John couldn't really hear them. Sometimes the chief would get on the radio and shout, *"Rescue, are you gonna circle all day, or are you gonna come to the fire?"*

John would stay in the kitchen all night, and that meant no one else had to do the house watch, which involved staying up and listening for radio calls. Gradually, the other guys also began to hang around in the kitchen all night too, until it became a Rescue 2 ritual for everyone to sleep downstairs. There were no beds there, just couches, but the guys made do. Sleeping in the kitchen saved you time getting on the rig and you never got left behind.

Every day tour he worked, John was sure to wash and polish the rig. At 9:15 AM, just after roll call, he would pull it out in the street, rinse it off with the hose, then whip out a can of Lemon Pledge and start polishing. He loved his job, and he was quick to challenge anyone who complained about the job or about Ray Downey. He couldn't stand to hear anyone bitching about the pay or griping about the hours. To John this was the best job on earth, period. John got along with everybody, even Gallagher's boys, but it was clear that he was Downey's man. Little by little, the guys who were on the fence about the new captain started to come around when they saw the loyalty Downey inspired. But what really turned them around was working a few fires with him. Some of Gallagher's men, like Richie Evers, went from hating him to being his biggest defenders.

When Ray Downey first arrived, Richie, like all the other guys, had his doubts about the new captain. But after some fires with him, where he saw how decisive and smart Downey was, Richie was sold. During Downey's tenure at Rescue 2, Richie got his nickname, Sergeant Evers, because he drilled the men like a marine drill instructor. Nevertheless the guys loved drilling with Richie, especially young gung ho firemen like Terry Hatton. Terry pestered Downey for months and months to get a spot in Rescue 2. The captain kept telling the kid to go out there and get more fire experience. Hatton would be back like clockwork, month after month, looking for a spot in the Rescue. One day Downey gave in. If Terry wanted the spot this bad, maybe he deserved it. Terry impressed everyone in the company with his desire to learn. Pete Bondy and Richie had never seen anyone like Terry in all their years at Rescue 2. Terry would go straight for the rig as soon as he got to work, and he'd learn how to use a new tool every night. The older guys only had to show him once, then Terry would practice what he had learned into the wee hours of the night, until he had mastered the tool.

The guys took their jobs seriously, especially in front of Downey. But they also knew how to have fun in the firehouse when the boss went upstairs. Mike Esposito, a wise-cracking muscle man, would do a stunt where he tumbled down the stairs. Bob LaRocco, a Tin House alum like Bob Galione, would always get tangled up in a wrestling match. But as soon as the door of the captain's office swung open, the firemen would instantly leap to attention, as if nothing were amiss.

All the Rescue 2 men loved to play jokes on Richie, because it was so satisfying to get his goat. Richie wore glasses in the firehouse, but it was impossible to wear them inside a fire because the mask didn't fit with his glasses on—plus they could melt. Without his glasses Richie saw only vague shapes and outlines, but it was enough for inside a fire, where you could hardly see anything anyway.

One evening, standing out on a sidewalk after a fire, Richie struck up a conversation with a young firefighter from Ladder 105. Richie didn't have his glasses on, because he'd left them on the rig. Watching him chattering away, the other guys from Rescue 2 broke out in gut-busting laughter when they saw what Richie didn't: the firefighter he was talking to was a woman. Richie was the kind of old-school guy who would never in a million years just strike up a conversation with a female firefighter. And yet, at job after job, he kept running into her without his glasses on and shooting the breeze. For weeks the other guys didn't tell him who he was talking to. Finally, they couldn't resist breaking it to Richie. "Hey Richie," one guy said. "Do you know who you were just talking to?"

"Yeah, some guy from 105."

"She's from 105, but she's not a guy."

"What? Get the fuck out of here." Richie ran back to the rescue rig, put on his glasses, and sprinted over to the woman in question. "She is, she is," he yelled back to the guys. He was more stunned than perturbed at his fellow Rescue 2 members. He appreciated a good joke and he had to hand it to them for that stunt.

More often than being the foil, Richie was the comedian. He had the sharp, outspoken sense of humor that all the guys loved. At another fire, a probie sheepishly approached the Rescue 2 guys and told them, "My officer sent me over here to see if you guys could treat this burn."

Richie stepped up and said, "Lemme see it."

The probie turned down his collar and revealed a small burn about the size of a quarter. Richie leaned over and planted a kiss on the kid's neck. "Now your boo-boo is all better," he said sarcastically. Then, gruffly, "Go back to your company."

RAY DOWNEY didn't open up to his men and lay his heart out on the kitchen table. That wasn't his style. He just told them exactly what they were going to do, and they did it. The rest of it, all the bullshit in the kitchen, what they might say behind his back, was unimportant to him. Protecting his men came first. Downey wanted discipline. He knew that his single most difficult job was not rescuing civilians; his biggest problem was the twenty-five kamikazes working with him. They were more than willing to risk their lives to save someone, but Downey was the one who would have to make the house call to their wife if one of his guys bought it. Downey didn't want to ever have to make that trip. So he needed to keep his men in line.

Only one guy in Rescue 2 was able to fool around with Downey and live to tell about it: Mike Esposito. Espo, as he was known, was a constant joker. When a mangy stray wandered into the firehouse, he took to calling it Dutch, the name of Downey's dog. That wasn't the kind of remark that too many guys would risk, but Espo could push all Downey's buttons and get just a chuckle in response. Espo was sometimes called the captain's illegitimate son because the guys figured that was the only way he would be allowed to say that stuff to Downey.

Bob "Rocky" LaRocco was a different story. He had gone with Bob Galione and Mike Pena, his buddies at the Tin House, to Rescue 2. A good tough fireman, Rocky had a habit of winding up in the wrong place at the wrong time. That in itself was no crime; it happened to everyone. But Rocky was always getting caught there by Captain Downey. One night, on the first job of the night shift, Rocky was assigned to head to the roof of a burning building, but he got waylaid on the first floor. Downey saw his roof man still on the ground level and screamed at him, "Next job, you go to the fucking roof, no matter what."

Rocky got the chance to redeem himself just a few hours later when Rescue 2 pulled up to a basement fire in an occupied apartment building. Downey and his can man went into the cellar. Rocky went around the back, heading to a fire escape that he planned to climb to the roof. In the backyard, however, a chief grabbed him and said, "We just got a report of a person trapped on the first floor. Go in and see if you can find 'em."

"But Chief, I gotta go . . . ," Rocky sputtered.

"Get in there now," the chief interrupted.

So Rocky headed into the first floor, searched it, and, finding no one, dropped down the cellar stairs to see if anyone was trapped there. Exploring the basement, Downey heard something on the stairs above. Shining his flashlight up, he saw Rocky, who peered back like a deer caught in the headlights. "This is not the roof!" Downey roared.

With an ordinary captain, all this control and attitude might have led to uncharitable speculations about whether the tough-guy act was just a way for the captain to make sure the men didn't upstage him in a fire. Not so with Downey, for he quickly proved to the men that he was as strong a fireman as they had ever seen.

One Sunday, the captain and John rode into work together, as usual. In addition to all the time these two guys spent in the rig, they also had two hours a day in the car together, commuting to and from work. They both liked to get there early—Downey to do paperwork and John to hang out in the kitchen. On this Sunday afternoon they arrived at 4 PM and found the other men out of quarters, on a run. Downey went up to his desk to start on a tall stack of paperwork. John poured himself a cup of coffee and sat down to read the paper. Not long afterward, Downey heard a call come in over the radio for an alarm box. The box was nearby, so Downey went to the window and glanced out. A plume of thick black smoke was already decorating the skyline. Downey and John jumped into the car with what gear they could find in the firehouse and sped toward the box.

By the time they got there Engine 210 was stretching a line. But down on the street, people were screaming "Baby Michael! Baby Michael!" People said baby Michael was trapped on the first floor. The place was filled with smoke; it wasn't vented yet because the ladder company hadn't arrived. Downey didn't even have the option of donning a mask, since his mask was on the rig, somewhere else in Brooklyn. But he probably wouldn't have worn it anyway. It got in the way; he didn't need it.

So Downey scaled the outside steps to the parlor floor of the brownstone. He hit the floor and crawled in, just as another fireman was leading an older man out. Downey pushed in to the rear, where he thought the bedroom would be. He scooted past the dining room.

Helmets record the experience of each Rescue 2 fireman. Many of these old leather helmets are still in use, burned to a crisp and flaking apart from decades of fire duty.
*(Jefferson Miller)*

Captain Phil Ruvolo *(left)* and his chauffeur, Firefighter Bob Galione, have a deep, unspoken communication. They trade laughs based on jokes never uttered out loud.
*(Jefferson Miller)*

Captain Phil Ruvolo has to make a constant and instantaneous calculation of risk vs. benefit. When should he let his men risk their lives, and when should he hold them back?
*(Jefferson Miller)*

In 1984 Captain Ray Downey moved Rescue 2 to 1472 Bergen Street in Bedford-Stuyvesant, Brooklyn.
(*Jefferson Miller*)

Mark Gregory (*left*) and Bob Galione at the kitchen table. Bob rules the kitchen like an Italian grandmother.
(*Jefferson Miller*)

John Napolitano (*left*) and Paul Somin in the early morning hours before the day tour begins.
(*Jefferson Miller*)

On his way to a fire, Lincoln Quappe was always chewing a gnarled cigar.
(*Jefferson Miller*)

From the time he set foot in Rescue 2, John Napolitano (*right*) was always eager to learn everything he could about rescue firefighting. He is seen here with Phil Ruvolo.
(*Jefferson Miller*)

Captain Fred Gallagher led Rescue 2 from 1973 to 1980, a period of intense fire duty. *(Courtesy of FDNY)*

Richie Evers came to Rescue 2 in the late seventies and stayed for twenty years, becoming the toughest veteran in the company. *(Steven Scher)*

Gallagher's men saw big jobs like this almost every night they worked. *(Steven Scher)*

Evers kneels to cut a roof in order to vent a blaze that burns below him. *(Warren Fuchs)*

Ray Downey, Rescue 2 captain from 1980 to 1994, initially proved himself to his men by his ability to breathe smoke all day. "The new captain can definitely take a feed," the Rescue 2 firemen said after they went to a few fires with Downey. *(Warren Fuchs)*

Downey helped make Rescue 2 into one of the toughest, most efficient, and most respected companies in New York. Gradually the house's reputation grew beyond the city, and Downey's crew became known as some of the most experienced firemen in the country.

*(Warren Fuchs)*

After rescuing "Baby Michael," who turned out to be a twenty-seven-year-old disabled man, Downey went to City Hall with his family to receive a medal. Pictured are Joe, Kathy, Marie, Ray, Rosalie, Ray Jr., and Chuck.

*(Warren Fuchs)*

Downey pushed hard to make scuba a part of the rescue companies. In the eighties, Rescues 1 and 2 were the first to be outfitted with scuba gear, and they sped to drownings, capsized boats, and marine emergencies throughout the city.

*(Warren Fuchs)*

Captain Phil Ruvolo gives quick instructions to his men outside a blaze before they plunge into the smoke and flames.
(*Jefferson Miller*)

To Bob Galione, a good job is three floors of fire, fifteen rooms going. When he stumbles out of a good fire Bob feels like he's just fought against a force of nature and won.
(*Jefferson Miller*)

The cubbies are where the men keep their fire gear, which always smells of sweat and smoke. The guys take a certain pride in the lingering stink of their stuff—it means they've been busy fighting fires.
(*Jefferson Miller*)

New recruit John Napolitano always sprinted down the street toward the fire faster than any of the old timers.
(*Jefferson Miller*)

Kevin O'Rourke with some neighborhood fans. When a kid came knocking on the firehouse door for air for his bike tires, Kevin wouldn't just lead him to the pump; he'd give the bike a ten-point checkup.
(*Rescue 2 Collection*)

On the afternoon of September 11, 2001, Captain Phil Ruvolo and
Lieutenant Pete Lund rest for a minute after being pulled off the pile
because Seven World Trade Center was about to collapse on top of them.

*(Ron Agam)*

Captain Ruvolo welcomes John Napolitano back to the firehouse after his vacation. The captain called John the "good son" because, after a slow start, he proved himself to be a dedicated member of Rescue 2.
(*Jefferson Miller*)

After a good fire, the men retire to the kitchen to brag, kvetch, and talk for hours. No one ever talks about the fear, that would be considered an infraction of the Rescue 2 code.
(*Jefferson Miller*)

The kitchen is the center of the firehouse universe. It's where justice gets meted out. Everyone is equal in the kitchen, from the new kid to the chief.
(*Jefferson Miller*)

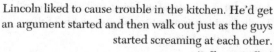

Lincoln liked to cause trouble in the kitchen. He'd get an argument started and then walk out just as the guys started screaming at each other.
(*Jefferson Miller*)

Flames licked out into the hallway. In the back room he felt a bed and, from his crouching position on the floor, reached up for the baby he hoped to find. He realized right away that this was no baby. Downey pulled a twenty-seven-year-old crippled man down to the floor and dragged him out the way he came. As he emerged people came up to him saying, "Baby Michael's alive!" Turns out the people on the block called this grown man "baby."

Downey could breathe smoke all day, a natural talent that became part of his legend. Sometimes Downey would be next to a man who was entangled in his mask and its hoses, trying to shout over the steady groan of the air supply. The guy would hear Downey bellowing an order to check that room to the left. No mask, no gasping for breath, just lungs. Back at the kitchen table, after Downey had retired up to his office, they'd say, "The captain can definitely take a feed." He also had the presence of a general on the battlefield; at big fires or disasters people looked to him for a plan of attack.

Downey wasn't big, tall, or muscular. He was just about average build, average height, and, though he was in good shape from long-distance running, he wasn't a he-man like Fred Gallagher. But he had penetrating blue eyes that could stare anyone down, and he was a master of the meaningful look. His firemen and his family made deciphering his expressions into an art form. A nod or a quick glance was his way of conveying not just the extremes of approval or disgust but also more subtle gradations of feeling. Inside a fire, what he brought, more than anything, were the lungs and the voice. Downey had a raspy way of speaking, with rough breaths punctuating his words. Magically, his voice was the perfect tone and timbre to cut through the din of a good fire. Eventually, his leadership skills would be utilized at fires and disasters all across the United States. But his way of leading men in disasters came from one place: hot, smoky hallways in the borough of Brooklyn,

By the time Downey's first year at Rescue 2 had passed, the men realized that he was the future—theirs and the company's. They knew they could count on the captain. Even though Downey was distant, when a Rescue 2 man needed help, real help, he got it, no questions asked. When one man's wife was sick, Downey arranged for other fire-fighters to cover his shifts while putting the man down as having worked his hours, so he didn't miss any paychecks. When Bob LaRocco, one of the captain's favorite whipping boys, was waking his deceased father, Downey was the first to show up at the funeral home.

Downey would enter a fire ahead of everyone so he could coach and direct the guys from the engine companies as they headed down the hallway. He had a knack for always getting everybody off the stairs just before they collapsed into the fire. There was never time for him to deliberate or to discuss; he just intuited the right moves. He was a natural. And he had a certain understated elegance to him too—like the time a building collapsed around them, and Downey simply said coolly, "Okay, guys, let's just walk out of here nice and slow," as if they were departing from a church picnic.

DURING THE DOWNEY YEARS at Rescue 2, the borough of Brooklyn was changing rapidly. In the fire-intensive seventies, Rescue 2 had been able to perform efficiently without being right in the middle of the ghetto. Throughout the Gallagher era, Rescue 2 had been situated in the middle-class black neighborhood of Fort Greene. Even after a night in the toughest, meanest parts of the borough, the men had a tranquil firehouse to return to. But after many years of sharing quarters with Engine 210, his officers decided that it was time for Rescue to get its own firehouse. There was tension with 210; the Rescue 2 guys often convinced the young bulls at Engine 210 to trans-fer to busier houses, where they'd get more fires. That didn't win friends at the engine. Also, the Rescue 2 guys ran around to fires so much that they rarely got to do much cooking or cleaning and happily mooched off of 210.

Downey found a vacant firehouse, which had once been the quar-ters of Engine 234, on Bergen between Troy and Schenectady. The house was nestled right between Crown Heights and Bedford-Stuyvesant, two of the most active fire locales in Brooklyn. The neigh-borhood surrounding the place was besieged by the crack epidemic, which was just hitting the streets full-force. Crack made the ghetto a meaner and nastier place than it had been before. Bergen Street, on the block where Rescue 2 would be located, became a hot spot for drug activity because there were no respectable buildings there. That meant no good citizens to complain about junkies on the stoop or dealers on the sidewalk. The crack years of the eighties brought extremes of inhumanity and made the rioting, arson, and rampant crime of the seventies look almost innocent. The looting and rioting in the blackout of 1977 at least expressed something—the anger that people felt about their lives. Crack didn't seem to express anything but absolute desperation and depravity. The Rescue 2 men witnessed people willing to do anything for a quick high.

In 1984 Downey formally moved the company from its comfortable, decades-old house to the new quarters on volatile Bergen Street. The Bergen Street firehouse was small and old, and the years had not been kind to the building. But the men of Rescue 2 united to fix up their new home. Most firemen endlessly puttered around at their private homes—building new decks, adding on rooms, fixing up basements, and so on. Now they had the chance to use these skills at work. They were in their element, discussing power tools and Sheetrock, mulling over blueprints and knocking down walls. Home Depot would be built as a temple to men like this. They spent all of their time, including their off-duty hours, making the ramshackle place into a home. One firefighter, Glen Harris, brought in a heating gun and started to strip off all the paint that had accumulated over the years, decades' worth of shoddy, good-enough-for-government-work paint jobs. Glen was at it day and night, trying to make the new Rescue 2 into a firehouse that was, if not exactly beautiful, at least not about to be condemned.

Everyone did his part. When the men went out on calls late at night, they would commandeer blue police parade barricades left out on the street. Then, back at the firehouse, they fashioned the sawhorses into shelving for the apparatus floor. One day a cop came in to use the bathroom and started scrutinizing the blue shelving, laughing as he read the "Property of the NYPD" markings. The guys built a shiny new bathroom next to the locker room. One fireman found an enormous wooden spool, eight feet in diameter, used by Con Edison for storing heavy-duty electrical wire, and somehow managed to truck it in to work—it made a perfect kitchen table. Larry Gray, the best artist in the company, was charged with carving a giant bulldog on the tabletop, and Mike Pena, a new guy, had to bring in gallons of marine resin to seal the tabletop against spills. In the kitchen, the men repointed the bricks. On the apparatus floor they removed the old fire pole to make more room for gear; everybody hated those things any-way—it was way too easy to pop an ankle shooting down the pole at four in the morning. They had to scavenge for most of the building materials they used since the city funds for remodeling weren't ade-quate for the job. Most of the stuff came from industrious members who spied a pile of two-by-fours or a sack of cement about to be thrown out, grabbed it, and tossed it in the back of their truck to take to the firehouse.

Theft was a huge problem on Bergen Street. On a few occasions, while the men ate dinner in the kitchen, thieves climbed the fence, dropped onto the roof, and robbed the lockers upstairs. When the company went to a fire, it was open season on the firehouse. After see-ing the truck disappear around the corner, a junkie would kick in a front door panel and have a small child wiggle inside to open the door, then enter and rob Rescue 2 blind. One night the guys came home to find a man and a woman wheeling the company's television set on a stretcher, a novel way to carry the loot. The firemen gave chase and pursued the two through the firehouse kitchen, out the back door, up to the fence. Once they reached the fence, the firemen stopped in their tracks and waited. The two thieves climbed furiously until they hit the barbed cyclone wire and got stuck in its prickly metal spikes. They weren't going anywhere. The Rescue 2 guys had to climb up, rescue the criminals, patch up their wounds, and call a squad car—and an ambulance—to take them away.

After a spate of break-ins at the house, Rescue 2 succeeded in get-ting an injured firefighter assigned twenty-four hours a day for secu-rity. When the men met the new guy, they launched right into their few ground rules for surviving at the firehouse. These included "Don't sit up in bed," one of the more important Rescue 2 safety principles, as the men had determined that if a bullet came in from the neighborhood park their best bet for escaping injury was to be lying down in bed. One new security detail got so scared when he heard rounds being fired (a nightly occurrence) that he rolled off the bed, crawled out into the locker room (which had no windows), and slept there. The officers' bathroom faced the street, and some officers would urinate from a sideways stance to avoid being targets in the window.

When Rescue 2 first moved into its new firehouse, the cops showed up every day, like clockwork, asking to borrow tools. "We need to hit a crackhouse. Can we borrow your bolt cutters?" they'd ask, and the rescue guys were happy to oblige. But before long the men of Rescue 2 noticed that these same cops were gathering late at night at the side of the firehouse to drink and then fire off their weapons. The firemen thought maybe this was all just run-of-the-mill off-duty fun, until they recognized some of the same police officers on the morning news in a report on a big police corruption case. Almost every cop in the precinct was eventually transferred out. Turns out the cops who borrowed the tools were robbing drug dens, then meeting next to the firehouse to divide the loot. They even had a radio code for the firehouse: *"Buddy Boy, Buddy Boy, 234,"* which meant Rescue 2's firehouse.

When the men finished up a night tour and went home in the morning, they routinely found the streets and sidewalk outside the firehouse littered with crack vials and the baseball fields decorated with shiny spent shells. The bustling neighborhood drug trade also meant that Rescue 2 had a new kind of fire to fight: blazes started by rival drug gangs or by criminals trying to rob dealers. Bodegas in the area were often drug fronts. In the eighties, even legitimate bodega owners needed extreme safety measures to survive. Clerks were stationed behind thick, bulletproof-glass walls in order to protect themselves from thieves. But inventive criminals soon found a way to circumvent the bulletproof barrier. They would enter with a can of gasoline, splash a little in the payment slot, then threaten to light the place up. When the Rescue 2 men traveled to a ripping bodega fire, they would often try to enter through a trap door from the basement that led to the area behind the counter. This way they could get in and save the sales clerk without having to get through the bulletproof glass.

One building right near the firehouse, just around the corner, on Troy and Pacific, housed an upright citizen, a landlord who was struggling to keep drug dealers out of his building. Rescue 2 went there twice in one week for water leaks, and each time the landlord told them about trouble with his drug-dealing tenants. Later that week, the police banged on the firehouse door and summoned the rescue guys to his building. They found the landlord lying in a pool of blood, the back of his head blown off by a bullet; there was nothing they could do to save him.

In spite of everything, Rescue 2 men loved the new firehouse for one reason and one reason alone: it put them closer to the action, closer to good old fire neighborhoods like Bed-Stuy and Bushwick, and to the new ones like Flatbush. When he saw how much quicker they got to fires, starting out so much closer to the work, Pete Bondy remarked wistfully, "If we'd been on Bergen Street back in the seventies, we'd all have been incinerated."

Pete was still superman at Rescue 2. Though he never boasted or bragged about his own achievements, everybody in the Brooklyn ghetto companies knew him. He didn't seem to age visibly, and he seemed to get even better as the years passed. If a stranger walked into the firehouse, this quiet guy sitting in the corner was the last person they would think was superman. But once they saw him at a fire they would understand. One of his famous tricks was to stuff his hood in his mouth, to filter the smoke, instead of using the mask, which would slow him down.

Yet even Superman was human. One day Pete was on the second floor of a big old commercial building, a large warehouse with rows of merchandise and piles of junk, and it finally happened: he got lost. Bondy had air left, and he wasn't afraid, but he just couldn't find his way out, no matter how hard he tried. When his air tank started getting low, he got on the radio and said calmly, "Rescue irons to Rescue officer, I'm lost up here. Can you guys come and get me?" The guys made the second floor and tied-off a search rope to guide them back to the exit. They found Pete after a short search and led him out. All the guys were quiet. Outside, next to the rig, John Barbagallo finally said what many of the men were thinking, "I hoped I'd never see the day when Pete Bondy got lost." If Bondy could get lost, anyone could get lost, run out of air, and die.

DOWNEY HANDPICKED HIS LIEUTENANTS at Rescue 2, the men who would command when he was off-duty. Downey also instituted a system whereby each of Rescue 2's officers had his own core team of men suited to his own approach to rescue firefighting. Ray Downey had a group of men he consistently worked with, as did each of his lieutenants. Downey's group was known as Downey's Clownies, and they did everything by the book. Downey saw to that. But his lieutenants had some unorthodox methods: Lieutenant John Vigiano's men, known as the SWAT group, were a bunch of truly addicted firefighters who listened to the fire and police scanners all day and night, trying to get the jump on jobs and make it in first. They organized "10-8 parties," so named because 10-8 was the radio code given when the company was out on the rig but waiting for work. On a typical 10-8 party, Vigiano and his men spent twenty-four hours straight just cruising the borough, never returning to the firehouse until their tour was over.

Vigiano's men would head toward every alarm that came in over the Brooklyn frequency even if they weren't assigned to it. If it was nothing they'd turn around. But, if it was a job, they might make it there before the other companies and get to do some first-due firefighting. Sometimes it almost seemed like being out there in the rig, just waiting for something to happen, actually created jobs. They'd be cruising through Greenpoint and catch a 10-75, then drift over to Flatbush, and find a fire there; it was almost as if the work was following them.

In the eighties the streets of Brooklyn were filled with abandoned vehicles, often stolen cars that had been trashed and left behind. Whenever Vigiano and his company passed one of these vehicles, his men would surround it with their Hurst tool, a powerful hydraulic cutter that could easily snap a seat post or pry open the roof. The guys would rip the auto apart, practicing for a pin job where they might have to free someone trapped in a car after a crash. It looked like something out of the movie *Bladerunner,* when at two in the morning this posse of men jumped out of the truck, descended on a burned-out car, and ripped it to shreds with their enormous, futuristic tools. For many years, Rescue 2 was the only company in the borough with the Hurst Tool, so they got to speed from neighborhood to neighborhood prying people out of wrecked cars.

During his stint as a lieutenant at Rescue 2, Vigiano also gave Downey a nickname that stuck. One day, a couple of out-of-towners were visiting the firehouse. Vigiano went around the dinner table, introducing each of the men. When he got to John Barbagallo, he said, "This man sits at the left hand of God." (Because he drove the captain, John sat to Downey's left in the cab.) That's when the guys started calling Downey "God." They always said it behind his back, never to his face. It began as a joke, but it became something more. Years down the line, after almost every man in the company had a story of being saved or protected by Ray Downey, the name seemed sacriligious, but appropriate.

Lieutenant Artie Connelly was one of the older officers at Rescue 2, so his group was known as the Alzheimer's group. Connelly had the shuffle of an old man. His legs seemed to move without his upper body. He was a stickler for "overhauling"—the process of searching out sparks and flames by nipping open ceilings and walls—and Connelly made sure that his men stripped buildings down to the studs. When fireman Mike Pena worked his first job with Connelly, he nearly passed out from exhaustion. Overhauling next to Connelly, he kept wondering where the other Rescue 2 firemen were until he opened a closet and found another fireman, Patty Brown, crouching down and hiding. "Close the door," Patty said. "Don't let him see me." As soon as they had to overhaul, the experienced guys would disappear because they knew that Connelly would work them into the ground. If there was a fire to fight, they'd be right there alongside him; but if it was overhauling, they didn't mind goldbricking a little.

One of Connelly's favorite training exercises was to scream into the back of the rig, over the intercom, that there was going to be a scuba drill. One day he called on the intercom to the back, and Tony Acato, a new Rescue 2 member, picked up. Connelly would not speak to new guys until they'd been in the company for six months. *"Put on Galione,"* Connelly bellowed. Then he ordered Bob, *"Tell that new guy he's got seven minutes to suit up for a scuba run."* Connelly directed his chauffeur to the Brooklyn Navy Yard. As soon as the rig stopped, Connelly jumped out the door and plunged into the water in full bunker gear. If the new guy was ready to dive in and save him, great. If not, Connelly would start to sink, which was powerful motivation for the rookie to dive in.

Connelly sometimes manned the Rescue 2 brand, one of the place's more outlandish rituals. The guys had come up with the notion of branding each company member as a way to test his tolerance for pain when they weren't inside a fire. They would preheat the industrial-size oven and then stick a crude metal Rescue 2 brand, homemade in the shop, into the heat. They'd wait till the metal was red-hot before branding the guys. Most men got branded when they were off duty so they could drink themselves almost unconscious to dull the pain. But even the biggest, toughest men felt the pain for weeks. Their arms were never pretty after that. Best-case scenario was a big ugly scar for life, in the vague shape of the letter *R* and the number 2. Worst-case scenario was an enormous, infected sore. The smell of burning flesh lingered in the kitchen for hours, and no matter how hard they scrubbed, a little bit of each man stayed behind on the metal tool.

BY THE TIME Rescue 2 had settled in its new home, Ray Downey observed something that only an FDNY member could have found dispiriting: the number of fires in Brooklyn—and everywhere else in the city—was going down. There were still enough fires to keep all the firemen in the ghetto happy, but compared to the seventies, things were *a lot* slower. Instead of taking this as bad news, Downey saw it as an opportunity. With more down time, he could build Rescue 2 into a crack emergency response team.

Fires are about instinct, knowledge, and experience; only rarely are they true puzzles. Normally they demand endurance and strength but not ingenuity. Not so emergencies—rope rescues, collapses, transit accidents. These are more varied and unpredictable. During the Downey era, emergencies became puzzles—or more accurately, brainstorming sessions for the Rescue 2 guys.

Most men had a specialty that they brought to the company—carpentry, elevators, or electrical work. If Downey found a victim stuck in a machine, there was usually someone on hand who knew how to start disassembling the machine. And there was always someone who was mechanically gifted and who could, at a glance, tell them what pins to pull and bolts to loosen to get the victim out. When the company faced a complicated emergency, Downey would huddle with his men and see what the guys came up with. These were the times for democracy; there was often someone in the huddle who knew more than he did about the specifics of cutting steel or breaking a lock. So this specialist was given the chance to use his knowledge to save a victim's arm or leg—or life. The Rescue 2 men loved the new challenge and thrived on using their minds to help save people.

After responding to a few hundred emergencies, Downey's men developed precise strategies for everything from impalements to people stuck in machines or pinned under subway cars. Downey brought in guys from the transit authority to teach the men about subway trains. After every building collapse Downey and his men went over floor plans, mapping out a strategy for the next one. They were always trying to learn more about what they might face.

The Lyle gun was a piece of equipment that appeared in the Rescue 2 logo but, before Downey's time, was seldom employed by the company. An invention that might have been created by Rube

Goldberg, the gun contained a powerful charge meant to propel a metal hook, attached to a rope, to the top of a burning building if a fireman was in trouble up there. The idea was that, if the hook is shot upward, the trapped fireman would be able to rappel down the rope to safety. Rescue 2 came up with an alternate function for the Lyle gun. Once Rescue 2 arrived on a scene to find a fire raging in a crack-house guarded by a frightening pit bull. They needed to get in to do a search, but the dog wouldn't let them enter. So one of the men improvised with the Lyle gun, removing its metal projectile and using the gunpowder charge to stun the dog. The story of their success with this technique quickly became a favorite around the kitchen table.

Once their appetite for solving mysteries was whetted, Downey's men became tireless scavengers, always on the lookout for unusual equipment they could use in emergencies. One day, while drilling at the Brooklyn Navy Yard, the men found a giant cargo net. One member suggested that it could be used to lift obese people whom Rescue 2 was sometimes called upon to help transport to the hospital. Enormously overweight people had long posed a problem for rescue firemen; they were all slippery flesh with nothing to get a grip on. With the cargo net, Rescue 2 could just wrap them up and lift. The men dubbed the new device the "hefty hauler," and before long they were getting special calls over the radio from chiefs in Manhattan who had heard about their ingenious new tool and needed help.

Once Downey opened the floodgates to weird emergencies, life got more interesting at Rescue 2. The company was once called to nearby Eastern Parkway to deal with a man wearing a ring that was cutting off his circulation. They actually carried a special tool on the rig for this eventuality. What the men didn't know until they arrived was that the ring was a cock ring. Only in New York. But the guys gritted their teeth and did their job. They whipped out their ring-cutting tool and cut the offending appliance off. That job was worth a solid week of laughter in the kitchen. And back in quarters, their dirty gloves finally got washed.

The new Rescue 2 also developed expertise in freeing people impaled on fences. In one thirteen-month period, they responded to five impalements. These jobs were both gruesome and emotional. In one case, a fourteen-year-old girl had fallen out the window of her Park Slope home. She fell from the second floor down onto an iron fence, landing sideways; the fence sliced through her kidney area. Two men from her family rushed out and held her up on the fence, while the women called the Fire Department. By the time Downey and his guys arrived, the ladder team was starting to cut the fence with their power saw. Rescue 2 leaped out of the rig, yelling at them to turn off the saw. It caused dangerous vibrations that could hurt the victim. Quickly the Rescue 2 men pulled out a cutting torch, a tool that could free the person with just a few quick cuts and without much heat. After some precision cutting, they carved out the area around the girl and lifted her onto a backboard with a piece of the fence still inside her body. They rushed her to the hospital with the metal intact, which prevented blood from gushing out of the wound. If the metal was removed anywhere but at a hospital, the victim might bleed out and die. Downey and another Rescue 2 guy went into the hospital operating room with their cutting tools, just in case they needed to cut the fence further.

The girl survived and a piece of that fence joined the collection of metal shards that were cleaned and hung on one wall of the firehouse, like trophies.

RAY DOWNEY recorded everything the company had learned about how to save people in desperate situations. His book, *The Rescue Company*, was a catalog of calamities, a long narrative of the worst fates that can befall people. Downey set about codifying the different ways people could get stuck, mangled, buried, and killed. And then he cataloged the many and varied methods Rescue 2 had come up with to save these people. Downey's book became the bible for rescue teams across the nation. In the eighties the FDNY bought a whole line of new equipment to help free people from horrible predicaments: tiny but powerful saws like the Wizzer, search cameras that could snake down into a collapse, ever more powerful hydraulic tools that could pry people out of heavy machinery. Downey always knew about all the latest tools and technology, and made sure that Rescue 2 carried everything and anything that might save a life or limb.

In one article Downey compared the way a rescue company in the seventies had freed a man's hand from a meat grinder to the way they would do it in the eighties using the Wizzer saw. Just a few months after the article appeared in the FDNY's official journal, Downey and his men got called to a butcher shop in Park Slope. It was an ancient place with enormous slabs of dark old wood, soaked with decades of meat juices, and an old-fashioned hamburger meat grinder that had been retrofitted with a motor. A young butcher's arm was in the grinder up to his bicep. His nerve endings had been constricted by the grinder, numbing his arm, so at least he wasn't screaming in pain. But he was plainly in bad shape. The Rescue 2 guys could see that most of his fingers had already been ground to a pulp and were now just a bloody puddle at the base of the grinder.

Downey saw his rivals, the Emergency Service Cops, on scene with the tool he had just written about. They must have read his article. The cops' Wizzer, however, was still in its original box, packed in Styrofoam pieces, which came cascading out as they prepared to use the saw. Downey ordered his men to pop out their Wizzer, which they had been training on, and they went to work on the grinder, first opening it from each side, then cutting out the corkscrew-shaped gear around which the man's hand was wrapped before sending him off in an ambulance.

The Wizzer incident was hardly the only example of competition between Downey's men and the police. Downey hated waiting for the police scuba unit to navigate the waters and get to a drowning person. Sometimes the cops got there fast and saved someone. But plenty of times the cops got there too late. Downey thought that, given the opportunity, Rescue 2 could get there faster, so he pushed hard to make scuba a part of the rescue companies. Rescue 1 in Manhattan and Rescue 2 in Brooklyn were equipped with diving gear and small, motorized inflatable boats. Suddenly, the rescue firemen were speeding to drownings and capsized boats, racing against the police. Everyone was pleased with this development except the cops, who had a growing rivalry with the FDNY. The "Battle of the Badges" between cops and firemen was starting to attract attention. One New York newspaper depicted FDNY and NYPD scuba divers dueling with knives above a drowning person, yelling at one another, "I was here first! No, I was here first!"

The first big successes for the scuba program came in Manhattan. Twice in one year a firefighter named Paul Hashagen dove into the waters off the island to pull people out from helicopter crashes. One woman, a traffic reporter, went down in two crashes in one year. Paul saved her the first time, but the second time they could not resuscitate her.

A little later, the action hit Brooklyn. One frigid winter night in 1985 a group of five teenagers stole a van and then went joyriding through the borough. The joy ended when they crashed into a fire alarm box and catapulted into Mill Basin Creek, at the southern end of the borough. Luckily, the alarm box went off when the van crashed into it. Warren Fuchs, who lived just across the street, was off duty and heard the alarm transmitted over the radio. He rushed to the scene and saw a van sinking into the water. Fuchs radioed the location to Downey, and minutes later, Downey and company were speeding through the borough. It was a long drive to Mill Basin, but the weather was working in their favor. In very cold water, some people can survive as long as forty-five minutes without oxygen. Chilly water splashed on the face induces the mammalian reflex, in which the body conserves a dwindling supply of oxygen and channels it only to the brain.

In the back of the rescue truck, Dave Van Vorst was suiting up for a dive, checking each piece of equipment and bouncing left and right as the rig careened through the streets. Richie Evers, a driver notoriously heavy on the pedals, chauffeured the truck. Richie drove the road race of his life that night. When the men piled off the rig, Fuchs was waiting with a witness who showed Downey where the van had plunged into the water. No air bubbles were visible.

Dave Van Vorst dove into the water, while another back-up diver stood by at the surface. It seemed like Van Vorst was down there forever. First he tried to force the van's side doors open, but the pressure made it impossible. Finally, he was able to pry open the back door. Using a search rope so he didn't get trapped in there himself, he explored the interior of the van. Reaching into the darkness, he felt a body and pulled out a girl. After untying his search rope, he shot up to the surface with the victim, whom he passed to shore, where the men started to revive her.

Van Vorst dove back down for more victims. Ultimately, he came up with all five people who went down. The other four couldn't be brought back to life; only the first girl survived. The incident was a powerful validation of the scuba program, and, two months later, the rescued girl visited the men on Bergen Street. Her only lasting problem was some numbness in her pinky finger.

EVEN WITH THE DECLINE in fires, Downey made sure his men had plenty to do throughout the eighties and nineties, responding to rescues and learning new skills. But they were still firemen at heart, and the months that stood out in the collective memory of Rescue 2 were months like January 1985, when the fires never seemed to stop. Rescue 2 guys called the cold-weather firefighting the "winter offensive," and that January's battles raged with an intensity that hadn't been seen since the seventies. When a shift reported into work, Billy Lake would ask the guys, "What battalion do you feel tonight?" and they would predict where the action would be. For that month, it was fire after fire, with only small breaks to eat and, rarely, to sleep.

Things got so busy that the guys coming in for the night tour would arrive at work, prep their gear, and hang out near the front of the firehouse, keeping watch for the returning rig. They changed shifts in the street because they were sure to get another run while they switched. Sometimes the guys from the new shift would be waiting, ready to change tours, as the rig glided down the street toward the firehouse, only to see it speed off to yet another job before they could even jump on.

In addition to the fires and the seemingly never-ending oddball jobs—meat grinders, machines, fat people, impalements—the company caught a number of structural collapses. One night in 1987, on Adelphi Street, Downey and his men were called to the collapse of an occupied four-story building. When they arrived, the place was a pile of rubble. Three Rescue 2 men worked from the bottom, while the other three worked from the top, cautiously removing pieces of debris and searching for anyone alive. After just a few minutes of digging, they heard a man's voice. Everyone stopped making noise, and then they yelled back. The man was stuck in a void, buried in debris, but he sounded okay. He complained about the pain in his leg, though he was otherwise in good shape. They dug around him, carefully trying to free him, and when they started to get close, the man called out, "I'm sorry for all the trouble I caused you."

"No sweat," Ray Downey told him, "When we get you out, we'll buy you a beer." A few hours later, they were at Kings' County Hospital, beers in hand, celebrating.

I T WAS 1989 when the Crown Heights riots erupted right on Rescue 2's doorstep. To Bob Galione, who had just come to the company, at first it seemed like a return to the chaotic days of the seventies. The city installed an emergency command post at Rescue 2's quarters, and the men were joined by ten engine companies, ten truck companies, and ten chiefs—basically all the people in the neighboring area that hated their guts. Still, the Rescue 2 guys wanted to at least appear civil, and firehouse etiquette dictates that you serve a hot meal to anyone posted in your firehouse. The problem for Bob, the chef, was that there was no way he could churn out chow for a 150 men in their tiny kitchen, even with an industrial-size oven and all six guys chopping veggies. The guys worried over what to do, until a team of caterers, hired by the city, invaded the firehouse, set up chafing dishes on the apparatus floor, and proceeded to serve a hot meal topped off by dessert and coffee and tea. That's when Bob and the other guys realized that it wasn't the seventies anymore, and they started calling the

Crown Heights riots "the catered riots." The catered riots turned out to be more of a series of loud shouting matches than a full-scale riot. Bob and the other guys suspected it wasn't going to be that bad when they saw some rioters half-heartedly tossing what looked like Molotov cocktails at a jewelry store on nearby Utica Avenue. The Molotov cocktails were in plastic soda bottles and just bounced off the front of the store, right back at the people who threw them.

When Downey's oldest son, Joe, became a fireman, he went to Brooklyn, where the action was. His first assignment was at an engine near Rescue 2, Engine 227. Joe commuted into work with his dad and John Barbagallo and listened to the two veterans talk about the job. This was Joe's main opportunity to learn from his dad, so he listened hard. Downey didn't talk shop with Joe much; he wanted him to learn about the job for himself. At one of his first real fires, Joe went to the floor above the fire and promptly fell into a hole that sent him waist deep into the floor, with his feet sticking out of the ceiling of the floor below him. When they drove home the next day, his dad and John started talking about the job. "Someone almost fell down on me," said Downey.

"That was me," Joe told him.

"That wouldda been great, my own son taking me down at a job."

Three years after Joe entered the FDNY, his younger brother Chuck came on the job. At Chuck's first job with his dad, he followed his officer and his dad into the blaze. Chuck made it through two rooms before he started gasping for breath and had to stop and put on his mask. His dad kept going, without air. For the life of him Chuck couldn't understand how his dad could breathe in there. When he asked the other men, they just shrugged and said, "The captain can take a feed."

Later, when they had some time in the department, Ray Downey sometimes couldn't stop talking to his sons about the job; it, and Rosalie, were the great passions of his life. But, in the early days, when Ray was still at Rescue 2, he remained tight-lipped about it to his boys, just as he always had been.

While Downey's sons were getting their feet wet, he was nearing the end of his stint at Rescue 2. He had stuck around for fourteen years, until he could truly call the company his own. He had moved the firehouse into the busiest fire district in the city, outfitted his men with technology that could save lives, and created a team that would follow him wherever he led them. As a captain with that much experience, he began to get treated as an authority; his opinion trumped rank.

Downey had become the man that chiefs trusted inside the fire. He had an uncanny ability to interpret all the small signs that pointed to imminent disaster. If he told everyone to get out, it would be asses and elbows to evacuate. When victims heard that raspy voice take charge, they knew they were safe. Downey and his men would often find people in the worst moments of their lives—hand halfway through a meat grinder, head impaled on an iron post, or chest pinned at the bottom of a massive collapse. It would be the captain's calming voice that got people through these disasters in one piece.

Although Downey became legendary in the FDNY for his ability to maintain a tough exterior in front of his men, there were a few incidents that even made Ray Downey crack. One of these occasions was a job where a probie, Kevin Kane, was trapped above the fire. When the men of Rescue 2 piled out of the rig, they could actually see the young man in a window a few floors above street level, on fire. The entire building was engulfed in flames, and the rescue men were forced to watch a fellow fireman suffer. There was nothing they could do but observe the tower ladder make its way up to that window, trying to get to the kid. Ray Downey, however, swung into action. Transmitting to Brooklyn on Rescue 2's boroughwide radio, he demanded that a helicopter be dispatched to medevac the firefighter to the burn center. When the tower ladder ultimately removed the trapped probie, he had burns on 80 percent of his body. Downey sent his best paramedic, Dave Van Vorst, in the helicopter. But the young fireman was too badly injured to save.

After that fire Downey called up one of his officers, Al Fuentes, and asked him to meet him after work at Potter's Pub, a firemen's bar on Long Island.

Downey and Al had first crossed paths on an emergency relief mission to Puerto Rico organized after a hurricane, to improve Mayor Ed Koch's chances with Hispanic voters. On their first day in Puerto Rico, Al, who was driving Downey around, backed the car into a rail and gently bumped the rear fender.

"What are you doing, fucko?" asked Downey.

But Al, an immigrant from Ecuador, had a fierce Latin temper that was a match for Downey's Irish ire. "Who the fuck do you think you're talking to?" Al snapped back. He was so mad that he gunned the car and drove them straight into the water. He stopped at the first wave and jumped out the door, yelling at Downey, "Now you drive me."

It might seem an unlikely beginning to a close friendship, but Downey respected men who stood up to him. He called up Al a few weeks after they came back from Puerto Rico and asked him to come to Rescue 2 as a lieutenant.

Now, at Potter's Pub, after the probie's death, Downey told Al how they tried to save the boy but couldn't. Al had never seen Downey so upset. The captain actually cried and the two men hugged. After a while, Ray Downey composed himself. That unreadable expression returned to his lined face, and he went back to being the leader his men relied on, the man who never cracked.

# SIX: DEPARTURES

NOT LONG AFTER THE ATLANTIC AVENUE FIRE, Bob Galione, who—as usual—is able to summon the energy and humor to help rally his brothers, christens Ruvolo "Captain Disaster." For some reason, during Ruvolo's time at Rescue 2, Bob and the captain seem to work the major fires and disasters but miss out on the constant stream of smaller but more enjoyable outings. Ruvolo, who normally appreciates Bob's humor, can think of nicknames he would rather have.

By now, Ruvolo has weathered the men's initial suspicions and the difficulties of his first weeks as Rescue 2's new captain. Ruvolo has taken the jokes and the jabs and the very obvious doubts about his fitness for the job. And at the end of it all, he emerged as their captain— not just by decree but by consensus, with the endorsement of the kitchen crowd. But the days after Atlantic Avenue pose a different kind of challenge for Ruvolo.

Atlantic Avenue will go down in Rescue 2 history as one of the landmark fires; it was an experience that changed everyone who worked it. Atlantic Avenue took a lot out of the men and they won't recover quickly. They are proud of helping to save Brian Baiker and Timmy Stackpole, but they are exhausted, and tormented by the question of whether the two men who died couldn't have been saved too. And again they are hit with the what-ifs: What if they hadn't been at that other fire but responding right out of quarters. What if they had found Blackmore right away and managed to do CPR in there? The questions are endless. When firemen die, they always are. Atlantic Avenue isn't the kind of job that brings the men into the kitchen the next day with stories to tell; it is the kind of job that stays on their minds when they are alone.

Two more firemen are dead—Scotty LaPiedra and Jimmy Blackmore. Everyone knew those guys. They had worked in Rescue 2's neck of the woods. The deaths hit the new men especially hard. Lincoln and John look older now; Ruvolo can see it in their eyes and in the way they carry themselves. They seem a little haunted, and more mature. At Atlantic Avenue they were face-to-face with something real and absolutely terrifying.

It's obvious to Ruvolo that Atlantic Avenue was the most important event he has seen as captain of Rescue 2. The next few months after a fatal fire are difficult. After Atlantic Avenue, as with all other fires where firemen die, the Fire Department offers the men of Rescue 2 psychological counseling. But most of the men are not the type to talk about their feelings, and, just as when Louis died, few are interested in the counseling. They have their own form of therapy: many hours around the kitchen table and some good jobs. To get through the recovery, the men have to rehash the night of the fire again and again, picking apart everybody's performance and replaying each moment. Maryann O'Rourke, Kevin's wife, thinks the amount of time spent going over each disastrous fire is a little crazy. She teases Kevin when she overhears what she considers the equivalent of postgame discussions during the phone calls he makes from home. But these talks help the men process what they have been through. They grieve, learn, express their fears, and prepare for the next time a fireman is in trouble. Everything, however, is indirect and subtle. No one says, "I was scared to death when we went in there" or "I felt like crying when we found Jimmy Blackmore." The men around the table wouldn't know how to respond to that. But they convey their feelings in other ways.

They obsess about the practical details of the job—who did what, who was screaming, who was calm, and who offered the best, most clearly thought-out suggestions when things got crazy. They learn who is unflappable under extreme duress and who is vulnerable. They pass on specific tips, making sure, for example, that in future fires someone will always head under the floor where they are working to see if there's a danger of further collapse. (Dave Van Vorst had done that successfully at Atlantic Avenue.) They reemphasize the importance of clearing all the way down to a victim's toes when trying to free him, as Bob had to do to get Baiker out of the rubble. There at the table, they create the stories that will be repeated for years to come.

If Phil Ruvolo had been at the Rescue longer, if he had more seniority as captain there, he would put in for medals for his guys after Atlantic Avenue. But as a new captain, he doesn't want to piss people off by making them think he's looking for too much attention. So he decides that it's best not to put in for any hardware. One of the chiefs on duty at Atlantic Avenue, however, calls him up and tells him he has to put in for a unit citation—an award for the whole company for its work at that fire. So Ruvolo, after some hesitation, brings the matter up with Ray Downey, who tells him that, while he should always try to avoid offending others, he should also be fair to his company and make sure that no one else claims the recognition Rescue 2 deserves.

There has always been a resentment and distrust of medals in the FDNY, especially among the guys from the engine companies, who are rarely rewarded for their bravery. Ruvolo's own father, who served in a Brooklyn engine company, once found a victim inside a fire and dragged him out, saving his life. After it was over the chief came up to Ruvolo's dad and said, "Great job, Phil. I'd like to write you up for a medal, but I can't write an engine man up for this." The problem was that finding victims is supposed to be a truckie's job. That's why Medal Day is sometimes called "Truck Appreciation Day" and why a good truck company usually has about thirty medals in its history while a good engine company has only four or five. But the medal situation is hardly Ruvolo's most important concern.

A few weeks after the Atlantic Avenue Fire, Dave Van Vorst, one of the most senior men in Rescue 2, wants a private word in Ruvolo's

THE LAST MEN OUT

office. Dave had been right there alongside the captain at Atlantic Avenue. He had helped to pull out the three men and get them to the ambulances. A quiet, strong, and absolutely dependable fireman, he is the bedrock of Rescue 2, a veteran with no liabilities, enormous experience, and admirable leadership qualities. But he tells Ruvolo, "I've pulled too many guys out of fires lately. I don't want to do it again." And his wife wants him to retire.

"Give it a couple of months," the captain says to Dave. "See if you still feel this way, and if you do, we'll put in the paperwork."

Dave agrees to wait a couple of months, and Ruvolo is glad. He hopes there won't be any more firemen to pull out for a long time. The captain needs this guy to stick around; it is impossible to find a young fireman with anything close to his experience. Dave has so much to pass on to kids like John Napolitano and Lincoln Quappe.

Still, Ruvolo knows how bad it must be for Dave if he's actually considering retiring. He knows what it cost the guy even to have the conversation. Ruvolo understands how he feels; a bad job can leave a man feeling old, spent, and unable to believe he can muster the luck to walk out of another one alive. When Ruvolo was a young kid on the job, he had yearned for action. Now, after Atlantic Avenue, he's ready for some time without any firemen or civilians to save. He's ready for some days in the kitchen without a tragedy to discuss or a funeral to dress up for. It's going to take a lot of days like that to get over Atlantic Avenue.

On December 18, 1998, a few months after the captain's conversation with Dave Van Vorst, Rescue 2 is called to a fire in the middle of the night. Ruvolo is off duty. On this occasion, Eddie D'Atri, an officer from Squad 1, is covering at Rescue 2. The action begins as a 10-75 on Vandalia Avenue, south of East New York. As recently as the sixties this area had goat farms and farm stands with fresh eggs for sale. But, since the seventies, it's been home to Starret City, a gargantuan federal housing project, and some other state-run homes for senior citizens and the mentally ill. Tonight's job is at a

169

massive high-rise senior citizen's complex. Rescue 2's men get some details on the radio as they swoop through the deserted city streets, speeding to the southeast of the borough.

When Rescue 2 pulls up and scopes out the fire, it looks dangerous. Fire is ripping through a twelfth-floor apartment. The apartment windows glow red. In concrete high-rise buildings like this one, fires rarely spread to adjacent apartments; instead they burn very hot in a contained space. What worries everyone looking up at the fire is that it's glowing red, but no flames are shooting out of the windows. The men can see that the windows are shattered. The fire should be blazing out of these window frames, but no flames are coming out. That might mean dangerous air currents at work up there. Air currents are always a problem in high-rise fires.

As the Rescue 2 men enter the building, they hear other firemen screaming for help on the radio. "Go, go, go!" yells Lieutenant Eddie D'Atri as they sprint up to the twelfth floor, flight after flight.

A series of frantic Maydays are coming over the radio. Every guy feels it in his gut: this one could be as bad as Atlantic Avenue. The other firemen are trapped and burning, screaming for help. Then the transmissions abruptly cease.

On the twelfth floor, Mark Gregory scopes out the hallway. He sees flames shooting out horizontally from the involved apartment, as if there's a huge blowtorch aimed out the door. Mark's heart sinks. One of the key strategies in a fire of this type is to control the entrance. Steel doors like the ones in this building can contain almost anything if they're closed. But if they're left open, the fire gets air and there's a chance it will spread farther. When a fireman first enters a fire apartment, he keeps a close hold on the door, ready to slam it shut and contain the fire if things get bad. But here, the firemen have lost control of the door. Now the men outside the apartment have no way to close off the apartment and contain the flames.

The guys from Engine 290, some of the best engine men on the job, are struggling to move the hose line in. Sizing up the situation, Mark and those with him suspect that the firemen in trouble are on the other side of the flames, in the hallway or the apartment. But there's no way to tell whether they're still alive. It's ferociously hot up here, too hot to stay for even a second without a hose line cooling things off.

The one hose line on the floor isn't giving them enough water to make it down the hallway. The engine men have to bail out, return to the stairwell, and regroup. When the Rescue 2 guys see 290 having to regroup, they know the situation is really bad. Dom Carino, on the nozzle, is one of the best nozzle men in the city. If he can't get the line in, nobody can.

At the opposite end of the hallway is Rescue 2 chauffeur Larry Senzel, an incredibly tough senior man who has gotten up to the twelfth floor from the stairwell at the opposite end of the building. Larry is the kind of guy who never says a word, but he's exactly the man you want for a trip down a hallway too hot for human life. (Bob remembers a fight Larry had back at the Tin House. It went three rounds: one in the kitchen, one out near the rig, and a third after Larry saw that he needed stitches and decided that he wasn't going to the hospital alone.) If anyone can stand the pain tonight, he can. Larry approaches the burning door from the opposite side of the flames. He finds one injured fireman on the hallway floor and struggles to drag the guy out. This victim is terribly burned, and Larry doesn't think he has a chance to live.

At the other end of the hallway, back with Mark, the engine guys from 283 rig up another hose line by tapping into the standpipe from the floor below and stretching the line up to the fire floor. With two lines blasting, the Rescue 2, 290, and 283 men make a push down the superheated hallway. They crawl on the floor into the flames, measuring their progress in inches. Both hoses are pumping at full blast.

Finally, they are able to make the turn into the fire apartment and chase down the fire. Just inside the door they find two more firemen, burned to death on the floor, with no hope of being brought back to life. It looks like these men opened the door of the fire apartment, not realizing they were creating a wind tunnel that would kill them. As the nozzle men finish the job, the other men solemnly pass the bodies out, taking off their helmets and paying their respects.

Rescue 2's men return to the firehouse shell-shocked. At least at Atlantic Avenue they'd pulled two guys out alive. Tonight no one caught in the blaze survived.

This kind of high-rise fire is much more common in Manhattan. Vandalia Avenue is close to the ocean, and the winds that whipped in off the sea are to blame for the deaths. The firemen who died followed standard procedure trying to get into the apartment. But the wind whipping through the high-rise was just too strong for the men. In the future, when firemen confront a wind-driven fire like that, up high where the wind blows so strongly, they will know to try to find another way to vent the blaze, making sure that they control the door. With all three men who entered the apartment now dead, it is impossible to tell exactly what got them killed. Which just confirms what the Rescue 2 men have known since Louis died: any little thing can get you killed on this job.

In THE PAST EIGHT MONTHS, six FDNY firemen have died, including two at the Atlantic Avenue Fire and three at Vandalia Avenue. Now Dave Van Vorst comes once more to Ruvolo's office. The look on Dave's face says that there will be no dissuading him this time. The captain just nods and puts in Dave's retirement papers. Ruvolo is sad to see him go. But a man can only take so many brothers dying.

This departure, however, is only the first that Rescue 2 will have to bear. Terry Coyle is also getting out. He has had a hard time recovering from snapneck. Now it's official: he needs surgery to fuse fractured vertebrae in his neck. Richie Evers is due to put in his papers too. Years of taking a feed have caught up with him. His lungs just won't carry him into fires anymore. If you retire after twenty years, you continue to receive half-pay; but if you go out from an injury, you get three-quarters pay for life. When the doctors listen to Richie's chest and look at an X-ray of his lungs, they confirm that he gets three-quarters. He's definitely earned it.

Richie's retirement marks the passing of an era. He is the last fireman in the firehouse who fought fires for all of Brooklyn's busiest decade, the seventies. In his twenty years at Rescue 2 Richie has become the heart and soul of the place, the definition of an old-school rescue fireman. Now, who will give the silent treatment to the new men? (Richie still has not spoken to John Napolitano.) And who will take his place as the tough old SOB who categorically refuses to work a double shift?

When Richie joined Rescue 2 in the late seventies, nobody in the Rescue worked twenty-four hours straight. It just wasn't safe when a single shift often brought so many fires. One night tour knocked you out completely for the entire next day. You were lucky if you could drag your ass home to bed, never mind work another tour. Richie is the kind of guy who doesn't give up the old ways—not ten years later, not twenty years later. By the time he departs, everyone in the firehouse is working twenty-fours except Richie and the guy opposite him, who has no choice in the matter. Billy Lake, who has worked this spot for years, has had to tolerate Richie's whims. Once Billy had to beg Richie for a twenty-four. Billy's son was playing baseball, and he needed to be there and cheer him on. He sheepishly asked Richie if they could trade single shifts and make a mutual: twenty-four hours straight instead of a night tour and a day tour. Normally, there was no way Richie would even consider the request. But Richie was godfather to Billy's son, and he couldn't say no. Billy worked the twenty-four, and on the tour where Richie should have been working, but had given his shift away, Billy caught two first-due jobs, a third alarm, and a second alarm—one of the best tours in recent memory. At the very end of the tour Billy just shook his head and said to the guys he was working with, "Wait'll Richie comes in."

Richie stormed into the kitchen and got right in Billy's face. "Don't you ever ask me for a twenty-four again, you motherfucker."

No one will ever be able to chastise covering lieutenants the way that Richie did. It became his personal art form. At one ripping fire in Manhattan's Hanover Square, the company was led by Eddie D'Atri, the same covering lieutenant who had worked the fatal fire on Vandalia Avenue. When Rescue 2 got up to the fire floor, a line was just lying there, waiting for the Rescue 2 guys to take it. D'Atri grabbed the hose line and attacked the fire. Paul Somin, his can man, backed him up. As they crawled down the hallway toward the flames, Paul stumbled over a body—a lieutenant who had passed out in the fire. Paul passed the officer to the men behind him for first aid. Then he and D'Atri went at the blaze anew. It was a tremendous amount of fire, but D'Atri wouldn't give up the nozzle until he had put it out. In the process, he got so badly burned that he had to go the burn center for emergency treatment. The Rescue 2 guys bundled him up in the stretcher and escorted him out to the ambulance. Richie helped them along but made sure to tell any passersby, "He's a covering lieutenant." He wanted the other firemen to keep thinking that Rescue 2 was invincible.

In one classic Richie story, a Rescue 2 lieutenant has just taken a horrible pounding at a fire and come out to lie down on the ground, overcome with exhaustion and smoke poisoning. Somebody fetches the oxygen tank and starts to give the officer oxygen. When Richie comes back to the rig and sees this, he gets furious. "Lieu, if you have to do that, go in the rig," Richie hisses. "You can't be out here letting people see that."

Ruvolo will miss having Sergeant Evers around to spend long hours teaching new men the tools. Richie was the veteran who would call a drill at 11 PM to keep the new guys in line and make sure they were learning every minute. He embodied the Rescue 2 ethos: fire and fun. Whenever Richie walked out of a job, he had a big smile on his face, even if he was beat up and burned up, out of breath and choking. It was his way of saying "fuck you" to the fire and honoring the endeavor. Even when he was pushing fifty-five and suffering from more medical problems than he could keep track of, after every job Richie would still tell new members, "Don't ever grow up."

Rescue 2 guys like to say that there are three ways to get out of the company: get promoted, get killed, or retire. Guys who came on young can retire in their early forties and move on to a second career or live off their half-pay. But the perks of retirement don't compensate for the losses: When a fireman retires, much of his connection to the firehouse disappears. As soon as a man leaves, the jokes, stories, and characters change. If a retiree comes by for a visit, he finds he's out of step with the rest of the men. He doesn't understand all the new nicknames and innuendo. And if he's visiting the firehouse when a call comes in, he can't just put on his gear and journey inside the flames with his buddies.

Richie is going to miss all of that. But he has no choice. He has to retire. He and his wife, Denise, have planned to move to Florida and set up a big guesthouse in their backyard, called the Rescue 2 Resort, where Rescue 2 guys and their families can come down for a visit anytime.

Ruvolo has suffered as much from Richie's sharp tongue as anyone. But he knows that Richie will be impossible to replace. Between sucking smoke, taking cinder blocks on the head, falling down stairs, and getting burned, the job takes a huge physical toll. It's a young man's game. And yet the young guys, often perfect physical specimens, just don't have the accumulated wisdom and experience of the guys like Richie Evers. It is one of the many ironies of the job that a man pushing fifty and a little out of shape who has lived through the war years is often a better fireman than a twenty-two-year-old who can bench four hundred pounds.

THE COMPANY HAS LOST three veterans, and the new captain's job has just gotten harder. Most of the Rescue 2 firemen who had doubts about Ruvolo's ability to lead, or who, at first, didn't trust his skills inside a fire, have been silenced by his work at the Atlantic Avenue Fire. That doesn't mean everyone likes Ruvolo, who, like most Rescue 2 men, has a strong and polarizing personality. But most of the guys feel they can work with him, and there is a grudging respect the captain can now feel. Still, this is the kind of closed society that takes years, not months, to infiltrate. Ruvolo can't let down his guard for a second.

That spring, at a fire in Bushwick, Ruvolo finds himself working with Timmy Higgins, one of the senior guys in Rescue 2. Timmy is a guy in the Richie Evers mold who loves to work hard and have fun. He and his brothers, also firemen, are the type of guys who might be belting out old Irish songs one minute and jumping into a heated fist-fight the next.

Timmy is on the lieutenant's list, awaiting promotion, so he'll be gone from Rescue 2 soon; he's another guy Ruvolo will hate to see go. But unlike the retired guys he's losing, at least Timmy can come back as a lieutenant.

Tonight Ruvolo and his men are at a fire in a three-story vacant building, and they've been sent to the building next to the fire to check the spread of the blaze. Ruvolo and Timmy are working the top floor of the building. Ruvolo is in the front room, and Timmy—in the back—stands up on a chair and opens up the ceiling with his Halligan, checking to see if flames have spread through the cockloft. When the truck company officer sees Timmy up there, he first tells him to get down, then shoves him off the chair. If the guy was just a fireman, Timmy would have knocked him down the stairs, but you can't hit an officer.

There has been tension between Rescue 2 and most of the sur-rounding companies for decades. Back in the eighties a Rescue 2 fire-fighter came up with a slogan for the company that just about everybody loved: "Rescue 2, we're not here to make friends." Now the guys have the slogan emblazoned on a work T-shirt along with the line "Rescue 2, Hate U"—as in Hate University, the place where you learn how to hate. The T-shirts are a big hit with the men, but they immediately piss off every other company that sees them.

Outside the fire, Timmy comes up to tell Ruvolo what happened with the officer. Ruvolo doesn't respond directly. He just says casually, "Excuse me, guys; I gotta leave for a minute." Then he saunters over to the truck officer, and he tells him, "Hey, not for nothin'," adding a strategic pause, "but you can't touch my men." Then the situation escalates: first big words, then a lot of little ones. Firemen nearby jump in and break up the fight. The Rescue 2 men love every minute of it. They have never seen Ruvolo in a scuffle with anyone.

A deputy chief who knows Ruvolo says to him, "You know, Phil, maybe you guys should take up now," suggesting they leave the scene. As Ruvolo walks back to the rig, he realizes that his guys have been watching the whole incident. They followed him from the beginning, figuring that he was going to get in trouble.

Timmy glances over at Ruvolo and says dryly, "Welcome to Rescue 2."

Lincoln Quappe puts his newly honed welding skills to good use in a showdown with Timmy Higgins.

Lincoln is the most hyperactive fireman at Rescue 2. He stays up until three or four in the morning, nods off in the kitchen for a couple of hours, and then he's back up at five or six, working in the shop or eating leftovers from last night's dinner. In the shop he has an old black-and-white television with a serious vertical-hold problem, but it's good enough to watch *Power Rangers*, which Lincoln likes to wake up to. If he can't actually look at the show because he's working on something else, he blasts it at a volume loud enough to hear over the torch or grinder.

Higgins, still awaiting his promotion to lieutenant, is less of a morning person than Lincoln. After he's been awakened from a fitful sleep for the third or fourth time by the relentless *Power Rangers* theme music, he jumps out of bed in a huff and flicks off Lincoln's television with an attitude that suggests he will lose no further sleep to animation. Lincoln, however, is the last to be easily intimidated; he waits a few minutes until Timmy falls back to sleep, then switches the TV back on.

The back-and-forth continues every morning until Lincoln devises a solution. For days, he welds a secret creation that he unveils one morning with his usual flourish. It's a protective metal cage that encircles the television and protects it from Timmy's lunges. When Timmy comes out that morning after one *Power Rangers* anthem too many, he can't turn off the damn TV. Finally, he just yanks the whole cage off the wall where Lincoln had mounted it, and throws it down on the ground. So much for the protective cage.

A FEW MONTHS LATER, Timmy Higgins becomes the latest in a series of senior men to depart. With Dave Van Vorst, Richie Evers,

Terry Coyle, and now Timmy gone, Ruvolo is looking at a substantially younger house. His biggest problem in training his new guys is not their inexperience, though; it's what's happening to the city and the borough. By 1999, the full effects of Mayor Giuliani's get-tough-on-crime campaign are being felt all over the city. The man Ruvolo made fly is taking off against the bad guys. Criminal activity, poverty, insurance fraud, and general desperation—all the things that make for many fires—are no longer as common in New York as they once were. The crack epidemic has died down, and the city is calmer than anytime since the fifties. Even the neighborhood around Rescue 2 is changing. Only two kinds of automobiles were once visible around the firehouse: ancient, beat-up jalopies driven by working stiffs, and Mercedes and Lexuses belonging to local drug dealers. Now, in 1999, working-class people, prospering from the economic upswing in the city, are driving respectable used Hondas and Fords, and they feel safe. It's not like the firemen are ready to move their families into Bedford-Stuyvesant, but still the place isn't looking nearly as dangerous as it did a few years ago.

Rescue 2's neighbors are ecstatic about the improvements in their quality of life. But the guys in the firehouse aren't so thrilled. None of them wants people to stay poor. None of them wants to see drugs destroy kids. None of them wants to see houses burn down. At the same time, no Rescue 2 fireman wants to spend his whole day in the kitchen reading lad mags. No one wants to go to seed watching endless *Law and Order* reruns on A&E. What's the point of being a Rescue 2 fireman if there aren't any fires left to fight? And, Ruvolo has to ask himself, how will the young guys get the experience they need? Even if there isn't a heavy load of day-to-day fire duty, his men will surely be needed for the big catastrophes: the building collapses, the Maydays for firemen trapped, and all the other emergencies this city never stops throwing at them. When the big moment comes, how will they be ready if they've never gotten to fight more than a handful of fires a week?

It's true that technical knowledge can be taught in training sessions, even if there are few fires. What can't be taught, except inside a burning building or in the midst of an emergency, are the intangibles of the job, the bits of practical wisdom that have to be experienced to be learned. At Atlantic Avenue, Ruvolo learned that even a tiny piece of hose could pin a man. Now, faced with a similar situation, he will check for a hose line and clear it completely before trying to drag the victim out. Fighting fires is sometimes about knowing exactly *when* to do something—when to close a door behind you and when to leave it open so you can get out, when a building is going to hold out long enough to put out the blaze and when it's about to come down. You only learn that through experience.

Above and beyond the knowledge fires provide, they teach firemen how to take pain (though the brand helps a little). The greatest Rescue 2 firefighters have typically been men who can take a horrible beating inside a blaze and still keep fighting. The best firemen can feel their skin burning off, their bodies withering from the heat, their extremities getting scarred with burns, and still keep going if they have to. That kind of toughness can't be taught. A fireman needs to weather a lot of jobs to develop it.

Experience also brings good judgment, which is crucial to staying alive in a fire. Even a silly little fire can kill a man if he doesn't use good judgment. When Phil Ruvolo walks into a fire, the first thing he thinks is, *How do I get my guys out?* After twenty years in Brooklyn he knows the best escape routes from almost any structure in the borough. His men don't want to leave a fire until the last possible moment. Many times he feels like Mom dragging the kids away from the playground, but he has to do it. And, at that moment, knowing the fastest way out is a matter of life and death.

Only sometimes does Ruvolo understand why he pulls them out. Maybe it's the way the heat is intensifying. Or the way a certain kind of ceiling looks when it's about to crumble. Other times it's the combination of a lot of little things that he doesn't like. The color of a fire signaling that it's just about to roar, plus a sound that might be a crack in the floorboards. He compares each fire he encounters to the thousands of other fires he's seen over the years, and usually he makes the right call.

There are some decisions he can explain; other times it's just instinct that tells him it's time to go. There are things his body knows that he can't explain in words. This is the kind of knowledge that's most difficult to teach. But it's what makes the senior men the protectors of the other guys in Rescue 2.

So Ruvolo faces a tough challenge in the upcoming months and years: at the same time that he's losing his senior guys, he's also in the middle of the biggest slump in fires in two decades. On top of that, there's the grief and the added fear swirling around the house after two fires where firemen were lost. Ruvolo can't do anything about the grief or the retirements, nor can he magically up the number of fires in Brooklyn. He just does what he can; he works his ass off to to prepare the guys for anything they might face.

Ruvolo starts making rough sketches of a training ground he plans to put in the parking lot. At the top he envisions a small platform that the men can use for high-angle rope work. Below that will be a large cargo container, one of the giant ones unloaded from freight ships. In this container will be a confined-space maze, designed to simulate a rescue in a tunnel or the narrow corridor of a ship's hull. Next to that will be a simulated trench, a hole dug by construction workers and reinforced with wooden shores to keep dirt from filling in the hole. When these wooden supports give way, workers can get trapped in the trench. It's a delicate operation for a rescuer who has to get his victim out without falling in himself.

Ruvolo manages to find a free, used cargo container through Louis Valentino's dad, from the longshoremen's union. And one of the Rescue's oldest and most loyal buffs, Sandy Williams, who owns an upstate lumberyard, donates the wood and fittings they need to build out the container. Unsurprisingly, the FDNY is no help in a project like this. As usual, the men are on their own. But, like Gallagher and Downey, Ruvolo is determined to give something extra of himself to take the house farther—to keep its edge and its reputation.

After Ruvolo secures the materials, he throws out the idea for the training area. Soon everyone is hunched over a diagram of the thing, making suggestions and additions and thinking out loud. This project is important to Ruvolo because it brings the men together in a shared endeavor, because it gives them a chance to hone their building skills, and because it lets them practice for emergencies. When Ray Downey moved the guys to Bergen Street, they bonded with one another as they worked on renovating the house. Ruvolo is hoping for something like that to happen again. Each man does what he's best at. Lincoln spends his days welding the parts for the rope tower. John is an all-purpose assistant, ready to bang nails or drill holes. Each day they work, John and Lincoln learn a new tool. The captain still quizzes John every chance he can get about the obscure points of rope strength or building construction. Gradually, over the course of a summer, this training area grows and, pretty soon, Ruvolo has a new place from which to hang John upside down.

For YEARS Bob Galione has ruled the kitchen at Rescue 2. There is nobody who puts as much love and care into his meals as the man now seen as Ruvolo's no. 2. No one else is willing to drive the rig out to the end of Brooklyn to find perfect Italian sausage and fresh tagliatelle. He's the best cook, hands down. And he doesn't just do Italian; Bob is also adept at the firehouse standards: pork chops, sliced steak sandwiches, pot roast. The food hasn't changed much in the last twenty years. The default is still a heaping serving of meat and potatoes with the occasional overcooked frozen vegetable and a dessert like Entenmann's cake or some ice cream. At the end of the meal the guys still call out the price. "Six a man," they'll say, and everyone will toss in their money, making change from the bills on the table.

When Danny Libretti comes to talk to Phil Ruvolo about joining Rescue 2, Ruvolo is thrilled. Danny is the senior man at 103 Truck in East New York, and he wants to cap off his career with some fire duty at Rescue 2. He's a seasoned veteran who's been on the job since 1982. All that's great, but more important, he can really cook. Danny isn't just a skilled amateur; he's a pro chef. He went to a French cooking school in Manhattan, and, as his side job, he works under French masters at four-star restaurants like La Caravelle. Danny brings the Rescue 2 meals to a new level.

He can whip up a chocolate soufflé that makes off-duty guys skip dinner at home because they want to stay and sample it. As soon as

Danny starts cooking in the Rescue 2 kitchen, the guys are ready to purchase any supplies he wants. Bob can't believe it. For years he's been asking for the basics: good knives, a better mixer, a new cutting board. And everyone kept telling him, "What we've got is fine." Now Danny comes in with his fancy French cooking, and the guys are ready to buy anything.

Everyone sees the glimmer of resentment in Bob's eyes, although he is wise enough to deny it. Even so, once the firemen see his weakness, they just have to exploit it. One day, when Bob isn't working, a few guys go through the entire kitchen and relabel everything in there, from flour to ketchup, with labels printed on the upstairs computer printer. Now, instead of "Heinz," there is "Danny Libretti's Famous Ketchup"; instead of "Morton's," there is "Danny's Gourmet Salt." It's their way of taking the piss out of Bob, who has taken the piss out of many men in his time. Leading the charge is Mark Gregory, who sees a chance to avenge years of abuse. They wait for Bob to come into work the next day and see their handiwork. The goal in the kitchen is always the same: to push someone's buttons until he's ready to snap. No one is exempt, not senior guys like Bob, not the lieutenants or captain, not even a visiting chief. Bob opens the refrigerator door. "What the hell?" he asks—and then realizes it's all for his benefit.

Bob is making the transition from just one of the many Rescue 2 guys to one of the most senior men in the company. Now, with Richie, Dave, and Timmy gone, Bob and Billy Lake are the senior guys in the house. A senior man has the responsibility of teaching the young kids about the Rescue, not just about the tools and techniques but also about the spirit of the place and their code of behavior. "You never tap out unless you're seriously hurt," Bob tells Lincoln. "No going sick for sprained ankles or pulled hamstrings. You work the tour out unless you need to be in the hospital." To John he says, "No R&R. Other companies take it. We don't. *Ever.*" (R&R is the two-hour break that companies can take if they've been in a brutal job. It means they can go back to the firehouse and relax.) Rescue 2 has a long-standing tradition of never going off duty. No matter how beaten down and exhausted they are, they work. "Being a good Rescue 2 member," Bob is fond of telling new recruits, "means having two things: heart and balls. Everything else we can teach you."

Finding himself a senior man is a surprise for Bob. It's hard to imagine how his eighteen years as a firefighter have passed by so quickly. He never expected that the job would become such a vital part of his life. He had no idea it would turn out this way. He became a fireman by accident.

Bob Galione grew up in the sixties and seventies in Bergen Beach, a small enclave of neat houses on the south shore of Brooklyn. Bob and his teenage friends had the run of the city—as long as they came home for dinner. They wandered the borough as if it were the frontier. They went horseback riding at Carroll's Stable or took the bus to the beach in Rockaway. Sometimes he and his friends got out of dinner at home and took the subway all the way to Canal Street in Manhattan's Chinatown to have a meal at WoHop's. For a couple of dollars apiece they could stuff themselves on egg rolls and chicken chowmein.

In Bob's neighborhood many of the fathers were civil servants. Bob's dad was a city cop, but he never encouraged Bob to follow him. He knew his son too well. Bob had a quick temper, and his dad didn't want him to have a gun to back up his rage. Becoming a fireman was a fluke, really.

Bob only took the Fire Department test in 1978 because his brother needed Bob to pace him on the mile run. Bob ran every day; his brother didn't. The night before the written exam, Bob was out drinking; he stumbled home drunk at 4:30 AM. Soon after that, his dad was vigorously shaking him awake—"like he was shaking a bum on the subway," Bob always says when he tells this story. Bob stumbled into the testing room and took the exam. He was drunk for some of it, painfully hung over for the rest, and had to run out to the bathroom a couple of times. But he passed. After that unpromising debut, he's had eighteen years of pure bliss on the FDNY.

PHIL RUVOLO knows he can always count on Kevin O'Rourke in a fire. This fact is driven home at a job in Flatbush. It's a fire in a multiple dwelling that starts out ominously, with some radio reports that don't sound good. When the rig pulls up to the scene, Ruvolo can see that the companies are having trouble doing all the little things that help to fight a fire: opening up the roof, hooking up a backup line, positioning ladders at key windows. It's a top-floor fire, and an engine company is in there already, trying to knock the fire down as Rescue 2 enters the building. The chief in charge waves them on up the stairs, and they run three flights, hoping to get a piece of the action. Kevin has the roof assignment and is shimmying up the ladder on the outside of the building, carrying a heavy gas-powered saw to cut through the roof.

When Ruvolo gets to the floor below the fire, the engine company is rolling down the stairs, frantically evacuating from the fire. "The ceiling collapsed," the officer says to Ruvolo. Then comes the kicker: "One of my men is still up there."

Before he leads his interior team up to the fire floor, Ruvolo gets on the radio to Kevin and tells him, *"Rescue to rescue roof. I need a hole."* Kevin can detect the urgency in his voice. Ruvolo does some shouting and screaming in the firehouse, but at a job he always projects calm control. Still, guys who know him as well as Kevin does can tell when Ruvolo really needs something. He really needs this hole.

The fire floor is filled with smoke and heat. Ruvolo's best chance for getting to the man who's down is for Kevin to break through the roof and make a vent, clearing the smoke and flames from the fire floor so Ruvolo and the men with him can do a good search. Up on the roof, Kevin yanks the starter cable on his Partner saw. The roof is hot, but there's no hole. The roof man has got to know where to cut the hole—and how to position himself so he doesn't get cooked when the fire starts to blast up through the hole.

Kevin puts on his mask, pulls up his hood, and climbs out over the part of the roof that feels like it's right over the fire. Then he starts to cut. He turns his body and his face away because as soon as he punches through, flames and heat are going to shoot out. As he's completing the job, the hole starts to spit fire and Kevin's gear starts to melt off his body. First his flashlight strap gets mushy. Then his mask starts to heat up, and his air tank feels as if it's falling off. Kevin is wearing the full array of department-approved bunker gear, and he needs every bit of it for protection.

Meanwhile, down below, Kevin's hole is draining some of the smoke and fire from the upper story, enabling Ruvolo and his guys to find the missing fireman. He's lying on the floor, disoriented and badly burned but still holding the nozzle. They pull him out and gingerly hand him off to firemen waiting below, careful not to cause the man any extra pain.

A little while later, after the fire is knocked down and the guys are huddling together outside, Ruvolo sees Kevin dismantling his gear. His plastic heat-resistant face piece is melted into oblivion on one side, where it faced the fire, and his coat is crisp and browned. His helmet is severely cooked. But they couldn't have gotten in there without that hole in the roof. And Kevin is all right.

UNTIL THE MID-NINETIES, the FDNY gave each man a yearly allowance to buy fire gear. As a result, the more penny-pinching firefighters would wear helmets or coats until they fell apart, pocketing their allowances. Most firemen's standard attire was a long coat and high boots over blue jeans and a T-shirt. The boots would be rolled down to the knees until a good fire; then the men would unroll them to cover their thighs. For the truck companies, the boots were less important. The truck had to move fast, and they were out of the way before the engine started spraying water, so they didn't need the protection of the boots as much. Some truckies even wore flip-flops inside the fire.

In the last few years, an extensive and mandatory new wardrobe, dubbed bunker gear, has replaced this ad-hoc gear. To encourage men to maintain it, the FDNY hands out new coats and pants when old ones get worn out or burned up. The new gear adds about twenty pounds to the firefighter's already overloaded frame but promises much greater protection in a fire. It's worth the extra weight: it's the last line of defense between flesh and flame.

Some veterans still rebel against the tyranny of a mandatory uniform by refusing to pull up their hoods or button their jackets. But most young kids have been intimidated into suiting up properly both by the threat of censure and by some hair-raising films they see at the academy. Borrowing from a well-known genre typified by such high school driver's ed specials as *Blood Flows Red on the Highway,* these shock films show probies what happens when they don't wear all their gear.

The mandatory FDNY bunker gear has its pluses and minuses. A fireman can take a beating without getting badly burned. With the old gear Kevin would never have walked away from this one unharmed. The downside is that, engulfed in the high-tech gear, a fireman loses his ability to feel the heat of a fire until it gets so hot as to be dangerous, even deadly. Sometimes he has to pull down a glove and expose some skin just to get a sense of how hot it really is. If you're working hard and don't take this precaution, you might get hurt. When they were testing out new gear a few years back, the Rescue 2 guys got a coat with a built-in thermometer. When the temperature soared to a dangerous high, the coat would start to make a noise, to tell the fireman to get out. "Your coat's crying," they used to say when the alarm went off. That piece of gear didn't make the cut.

When Kevin turns in his face piece to get a replacement, the quartermaster is skeptical. He doesn't believe that Kevin was wearing it when it got burned like that. He says he's never seen a face piece in such bad shape on a guy who lived. But when he looks at the smile on Kevin's face and the sincerity radiating from his blue eyes, he just has to believe the guy. The quartermaster gives him a new face piece and files the old one away for posterity.

THAT WASN'T THE ONLY TIME Kevin, who seemed like a softie, had proven himself to be one of the toughest men in the company. Once, at an apartment fire in Brownsville, Kevin is checking the floor above the fire for victims. When suddenly he sees the holes in the floor below him, he says, "Watch the floor!" as he plunges through the gap in the floorboards himself. He falls through to his chest but hangs on to the beams; the fire below is cooking his lower body and he dangles like meat roasting over the flames. But he doesn't scream.

Later, back at the rig, after the guys have pulled Kevin out, the lieutenant makes a point of telling everyone, "Guys, Kevin didn't scream." That's the most important thing to all of them—a fireman never screams if he can help it. Anybody who does gets branded a coward or a hothead. Screaming is a sign that things are out of control, and for a good fireman, things are never out of control. If he gets caught in a bad spot, he takes a deep breath and thinks of how to get out. If he needs help, it's, "Hey, can you give me a hand," uttered without inflection, not a frantic, "Help!"

Kevin didn't scream and he didn't panic. Even afterward, on the way back to the firehouse, all he says is, "It got a little hairy in there. But the brothers were kind enough to lift me out of the hole."

BECAUSE OF THE RETIREMENTS, Ruvolo is preoccupied with the search for new Rescue 2 members. He has to be choosy and careful, because these men will represent the beginning of the Ruvolo company—his first recruits. There is no shortage of candidates. Sometimes it seems that everyone in Brooklyn is coming in to see the captain and angling for a spot. When these candidates come to the Rescue, they always bring a cake and always head upstairs for a private chat with the captain. Downstairs the guys eat the cake and, at the same time that they review the baked goods, also discuss the guy upstairs talking to Ruvolo. Everybody knows everybody else in the FDNY, and when a guy walks in, if the guys don't know him personally, they always know somebody who can give them the dirt on him. After the guy leaves and Ruvolo comes downstairs to sample some of the cake, he hears what the guys have to say about the candidate. He listens to the men in the kitchen, but the captain makes the final call. It's not open for discussion. This isn't a democracy.

When Ruvolo picks a guy for Rescue, he considers many things. First, what kind of fire experience does the guy have? He has to be from a busy house, usually a Brooklyn house, or he doesn't stand a chance. Then the captain tries to gauge how a candidate fits in with the rest of the company and what special skills he might bring. Jimmy Kiesling, one of Ruvolo's first Rescue 2 recruits, is an army reservist who is nuts about ropes and rifles. He's taken the army's sniper course, and he teaches rope work at the FDNY's rescue school. Plus he happens to be a little crazy, in an entirely positive way, the same way those guys were crazy at Ruvolo's first fire: Kiesling doesn't seem to show pain or fear. And Ruvolo's been down a lot of hallways with Kiesling, so he's a natural candidate for Rescue 2.

Two other new Rescue 2 men, Pete Romeo and Duane Wood, both come from East New York's Ladder 103, the house Fred Gallagher commanded before he came to Rescue 2. Ladder 103 has a reputation for being tough in a fire, and even tougher in the firehouse. Its location is so bad it makes Rescue 2's neighborhood look safe by comparison, and on more than one occasion someone relaxing on the john at 103 has ducked bullets flying through the bathroom window.

It's not hard to see why Ruvolo picked Duane "Woody" Wood, a pumped-up Brooklyn native whose physique is decorated with tattoos. Or why he chose Pete Romeo, even though he has a history of pretty complicated relationships with his former captains. Ruvolo figures that what passed for an oppositional attitude before will serve Pete well alongside Rescue 2's collection of wise asses and tough guys. Pete may be able to fill some of the personality quota that Richie's departure has left unfilled. Besides, his attitude is not unjustified; he's an awesome fireman. And the captain likes the guy; Pete is a great storyteller. With Pete and Bob at the kitchen table, Ruvolo never has to worry about any prolonged silences. They both smoke like chimneys and flick the ashes from their cigarettes onto the floor, though Pete's voice is even harsher and smokier than Bob's. Despite everything, Pete has a way with strangers. He makes friends everywhere the company goes. Whenever they leave him alone on the street for a second, they come back to find him in a conversation with somebody, whether it's an old lady or a traffic cop. The guys take to calling him "Deuce," after the character in the movie *Deuce Bigalow, Male Gigolo*.

Pete and Woody get a chance to prove their worth almost immediately after they arrive at Rescue 2. Ruvolo catches a job with both of them in Bushwick—a wood-frame building with a raging fire on the first floor that threatens to spread to the whole house. The smoke on the second floor will kill anybody who is not brought down immediately. Pete and Woody charge up the stairs to check the second floor. Ruvolo is down on the ground floor coaching the engine team battling the fire.

Upstairs, Woody crawls toward the front windows, which he will pop to vent the smoke and make the search easier. Even twelve inches above the floor it's really hot, so he stays low to the ground and reaches up with his tool to smash open one of the windows. It shatters, but he feels a warm piece of glass slide down between his thick glove and his coat and touch his skin. Smoke starts to suck out of the window, and Woody follows Romeo into the rest of the house, searching everywhere for a body and smashing windows wherever he finds them. His gloved hand feels wet, as if he got sprayed with a hose line, and then he realizes that his glove is rapidly filling up with blood. He should go down and get the wound bandaged up. But hell, they should also finish this search, and he can't leave Romeo here alone. So Woody sucks it up and scours the floor, the bed, the bathtub, for a victim. By the time he winds down the stairs to find Ruvolo, he's feeling faint and he can see that it's not just an ordinary cut. Each time his heart beats, it spurts a red geyser of blood. That means an arterial bleed.

Woody tracks down Ruvolo and tells him, "Cap, I gotta go to the hospital."

Ruvolo is about to ask what for, when he gets showered with fresh red blood. "Hey, you're spurting on me," he jokes. But when he looks more closely at the wound, he can see that Woody is in trouble.

At every 10-75, a team of paramedics stands by for firefighter or civilian casualties. When Woody stumbles away from Ruvolo, shooting blood from his wrist, he's surrounded by paramedics, who quickly lead him to a stretcher. To them, the wound is a lot more serious than it is to Woody. So serious that they call for a police escort to close off a route straight to the heavy trauma center at Bellevue Hospital. Police cars all over Brooklyn and Manhattan respond to the call. Traffic stops on the Williamsburg Bridge to make way for the ambulance. A lane of the FDR Drive is sealed off so they can shoot up to Bellevue. Woody is whisked away from the scene; word goes out to the commissioner and the mayor that they might have another dead firefighter.

The wound turns out not to be as serious as the paramedics first thought. Woody will fight fires again, in a few weeks. Ruvolo feels tremendous relief, but when he thinks back on the incident, he feels queasy. The men of Rescue 2 have precious little concern for their own safety, and now that's *his* problem. Because a captain's job is to help the men negotiate the balance between risking their lives to protect others and trying to protect themselves.

THE STATE STREET COLLAPSE, which occurs on July 11, 2000, starts with a massive gas explosion that takes down a house in Boerum Hill, a pleasant middle-class neighborhood of tree-lined streets and brownstones. Rescue 2 hits the road with Bob driving Ruvolo. Eddie Rall, Paul Somin, Billy Esposito, and Ray Smith are bouncing around in the back of the rig. When they pick up the transmissions from the battalion chief on the scene, they know they have a long night ahead of them. As Rescue 2 nears the site, Bob can smell gas. Pulling up, the men see an enormous pile of smoking rubble. They can't even tell how many stories the house might have been.

Huddling for a quick consultation, the chief and Ruvolo wait with worried faces for Con Ed to turn off the gas. Sending men in before the gas is off is too dangerous; there could be another explosion at any moment. After a few minutes, Con Ed gives the word, and the Rescue 2 guys are in. They carefully probe the rubble, searching for voids where people might still be alive. Nobody knows for sure who was inside when the place blew. Neighbors say that at least a couple of people are almost certainly under the pile. If anyone is still alive, they'd be huddled in a void—a pocket of space in the collapse. Bob and two other guys work from the bottom up; the other three men start from the top. The work is treacherous, and the men move with extreme care, shimmying into tiny spaces and calling out for victims. Using their portable lights, they search the site for bodies and feel around where they can't see. But the place remains eerily quiet. Nobody is responding to their calls. It feels like a death scene.

After a half hour, rescue companies arrive from other boroughs to help out. When Ray Downey gets there, he orders heavy machinery for picking off the big beams and debris that lie across the top of the pile. Bob and Paul Somin find an old coal shoot that leads down to the basement, and they slide down it to where they think people might be trapped—near the stairwell. But the rubble is so densely packed that they can't find anyone. After a few hours of crawling through the rubble, the guys become increasingly convinced that no one is alive in there. It seems almost impossible for somebody to have survived under this kind of destruction. On the other hand, it's possible to be trapped and survive for days. The hours pass slowly as the men, moving carefully and methodically, attempt to search all the voids.

When they need to rest for a little while, the guys sit down on the curb and drink some water. They splash some on their faces to wash the dust out of their eyes. Kevin O'Rourke is resting in front when he sees a couple of TV news cameramen who look spent. He goes to where the FDNY is giving out water and brings two bottles of water back to the news guys. The other Rescue 2 guys chuckle when they see Kevin doing this. That's Kevin, bringing some water to a reporter, rather than resting himself. It brings a little relief from the tragedy to laugh about that.

The next day, after twelve hours of searching, and after removing much of the debris with heavy cranes, Ray Downey watches his men uncover a foot. They recover the body of a man who had lived in the building. He's dead, every bone in his body crushed. His corpse is completely limp when they heft him onto the stretcher. Later they find two more victims buried in the rubble. All of them seem to have died instantly when the building came down.

EVEN AFTER A HERCULEAN EFFORT like working the
State Street collapse, most men will head home, take a quick shower
and grab a few hours' rest, and then report to their second jobs. For
financial reasons, almost everyone in the firehouse has to work a second job. Ruvolo, who does not himself take on a side job, would prefer that his men be able to rest on their days off, but he understands
the reality of the situation: The cost of living in the city is just too high
for a family to live on one fireman's salary. Since Rescue 2 is a firehouse made up of experienced firemen, the guys are older than in an
average house. There are no twenty-two-year-olds just out of probie
school looking to go out every night and party. Almost to a man, the
firemen of Rescue 2 are married and have kids. A few guys are
divorced, a few don't have families yet, but almost everyone is a
provider with worries about making ends meet. So they work on the
side, often finding employment together. When somebody finds a
good side job, he tells the rest of the guys, and anyone who needs
work joins up.

Carpentry, plumbing, electrical work, contracting, glasswork, and
sail making are a few of the hands-on professions through which the
Rescue 2 moonlighters supplement their city paychecks. Now and
then a guy gets a desirable job, like lifeguarding or bartending, and
gets waxed in the kitchen for it. In the summer, one posse of Rescue
2 guys works at the Boardy Barn, a huge outdoor bar in Hampton
Bays.

This is not the rich and fabulous Hamptons, the turf of Lizzie
Grubman, Billy Joel, and Martha Stewart, but the other lower-rent
Hamptons, where schoolteachers and young firemen get together.
Mark Gregory works security for the Barn's weekly Sunday bashes,
and he brings on Bob Galione, Paul Somin, and John Napolitano as
bouncers.

Every summer Sunday, those guys have the opportunity to see
carefree versions of their younger selves partying, picking up girls,
and enjoying all the pleasures of single life that now seem so distant.

Maybe they're a little nostalgic, but they're also just grateful for being able to squeeze out a few bucks in such a relaxed venue.

John Napolitano, with his "athletic" frame, is definitely the odd man out among the bouncers. Bob's a big guy who radiates authority. Paul is pumped up and used to work as a corrections officer in a Nassau County jail. John is rail thin and under six feet tall. Because the Barn has only double- and triple-XL shirts for the bouncers, John swims in his uniform. Shortly after John starts, one of the bouncers points to him and asks Paul, "Would you let that guy throw you out of a bar?"

But John attacks the job as he attacks everything else. If he has to fight with a big guy, he doesn't take him down from the front—he sneaks up behind him. If he can't knock a kid down, he'll jump on his back and subdue him until a bigger bouncer comes along. This cracks up the other Rescue 2 guys. "He's riding someone again," Bob says, and they go to help John. He tries so hard that they have to like the guy.

John is finally making the transition from being an absolute pariah to being just one of the guys. Since he proved himself under the extreme pressure of the Atlantic Avenue Fire, the other men trust him more. He can tell he's starting to fit in, because now people are hanging out with him away from the firehouse.

One day, when Ruvolo is drilling the men, he poses a question about how many pounds a particular sling can lift. Chuckling in the back, some of the old guys turn it into a joke and yell out, "Thousands of pounds, Cappy." But Ruvolo senses that John knows the correct answer. John, Ruvolo has noticed, *always* knows the right answers but he's often wary of speaking out and risking a round of jeers from the other men. So Ruvolo directs the question to John: "All right, thanks, guys. And now, to the *good* son. What's the answer?" At first the "good son" thing is a joke, but as time passes, the nickname sticks. John is the good son, the one who does his homework, in contrast to all the bad sons who would rather fuck around than answer Ruvolo's questions. And John, who isn't considered an expert in anything yet, has begun to build his reputation as a guy who knows all the facts and figures of rescue work. That is the kind of information that will serve him well if he ever gets promoted to lieutenant and has to lead a rescue team himself.

ON A SUNNY SPRING DAY, around midmorning, Rescue 2 is in the firehouse, just about to go for lunch. A computer-synthesized voice, the modern equivalent to the old bells, calls the men down to the rig, and they head out for a local box, a first alarm. "Where we goin'?" asks Ruvolo. When they make the left turn onto Schenectady, he can see gray smoke snaking up into the sky. "Toward that smoke, I guess," he says to Larry Senzel, who's driving him.

Turns out the first-due box is just five blocks away. Larry leans a little harder on the gas pedal, and the guys in the back hear Ruvolo yell back to them, "It's a job," exactly what they want to hear.

Usually, each Rescue 2 man has ample time to put on all his gear and make sure every button and snap is closed, but this time the guys in the back have to scramble. Kevin O'Rourke is OVM (outside ventilation man), which means he'll scale the fire escape and shatter all the windows he needs to. Ruvolo, John Napolitano, and Paul Somin are the forcible entry team. If they're the first truck company on scene, which doesn't happen often enough for their liking, they'll probably get to do some of the fun stuff, like pop a few doors. John is hefting a fire extinguisher. Paul carries a Halligan tool and an ax.

The scene is a corner brownstone building, where flames are shooting out of the parlor-floor windows. When Ruvolo sees the flames, he gets on the radio and announces, *"Rescue 2 to Brooklyn, 10-75 the box,"* just as they glide to a halt. There's no drama in his voice at all—it's deadpan, a whisper, a Bogart imitation.

Larry screeches to a halt just across the corner from the fire. (The engines and ladders need to be close, so he leaves space for them.) Ruvolo trails his forcible entry team up the steps. They can see fire shooting out of the rear windows on the parlor floor.

"Let's go," says Ruvolo, as he looks through the windows next to the door and sees more smoke and fire filling the apartment.

With a quick strike of the ax, they're in. Inside, Paul Somin kneels near the door to the fire apartment and positions his Halligan tool in the doorjamb. John Napolitano puts down his can and swings the ax, with the blunt side striking the Halligan tool that's wedged near the lock. The apartment door starts to give way.

"Cap, hold the doorknob," says Paul. That will keep it in place for the next blow. "Johnny," Paul says, "you gotta give me one more shot." John cocks the ax again and lets go, this time driving open the door. Smoke starts to seep out of the entrance. The men storm the fire apartment in search of victims. Ruvolo hits the floor and makes for the front bay window. Reaching up with his metal tool, he takes out each pane, one by one. The shattering sounds on the street are fantastic.

Part of the fun of a job like this is the destruction. Anyone who ever wanted to smash a house to pieces, knock down doors, or break windows can understand the pure pleasure of ripping the place apart. Outside, on the front steps, Kevin O'Rourke, who loves to talk on the radio, transmits a message to the engine company, *"We've got fire on the parlor floor."*

Inside, Ruvolo, John, and Paul carefully circle the fire room, looking for victims. They can see nothing near the fire. The thick pea-soup smoke makes it darker than the darkest night. For Ruvolo, everything is narrowed down to what's going on between his two hands. The world around him ceases to exist; all he knows is what he can feel in front of him. Inside a blaze like this, a firefighter six inches away is invisible.

The smoke in the room tells Ruvolo something. When he shatters the windows, he wants to see the smoke ease outside. Often the smoke will be pulsing at the window, seeping out of any little leaks in the pane. If it doesn't move out immediately, it means that things aren't venting right. Ruvolo can really feel the heat as he moves toward the back of the apartment. The engine men are coming in behind him. "It's lightin' up," he yells to them. "Let's go."

The radio outside announces the fire. *"We're using all hands. It's a four-story MD [multiple dwelling] with fire out the rear."*

Ruvolo can't see the other members of his team, but he's never alone. He constantly shouts back and forth to his men.

Ruvolo guides the engine guys into the flames. "Let's go, 214," he says, and then maneuvers himself behind the nozzle man so he doesn't get cooked from the steam hissing off of the fire. "Get some water on it, let's go," Ruvolo barks. Almost a dozen men huddle near the hose line. The engine wants to be as close to the fire as possible before opening the nozzle. A hissing sound precedes the flow of water. When the spray opens up, it starts to knock back the flames.

"Lighten up on the line" Paul Somin yells to get help moving hose. The engine company moves toward the heart of the fire, carefully guiding the nozzle across the floor, up the wall, and onto the ceiling. The hose is way too strong for one man. Many old hands will tell you that the backup man, the guy behind the nozzle, is even more important than the nozzle man. The backup man has to anticipate the nozzle man's movements. Right now he strains to hump the hose into place. "Good job, boys," Paul says.

As the engine guys finish off the flames, Ruvolo and his men look for fire hiding in walls or ceilings. Fire can hide behind a wall or spread through a ceiling, and Rescue 2's job now is to rip the place apart and make sure there's no hidden fire still burning. Ruvolo heads down the interior stairs to the basement and starts to tear apart the ceiling. He finds a little fire up there and radios for the engine to bring the hose.

After they knock that down, Ruvolo radios to the chief, *"Primary search is negative."* That means they've tossed the place once and found no one. Now there's a secondary search for good measure. After a secondary search and an overhaul of the building, Ruvolo and his guys climb down the stairs and regroup in front.

"I went for the front windows," Ruvolo tells Paul. "Then I tried to follow you guys down the hallway, but it was lightin' up."

The chief comes over to Ruvolo. "Did you see any people leaving the building, anybody in there?"

"No, we forced the stoop door," he tells the chief.

Ruvolo walks to the rig, tosses his tool and helmet in the cab, and then struts around to the back of the rig to put on a fresh bottle of air. He straps in a new one, then pushes the gauge three times to reset it.

"It was a good first-due job, you know," Ruvolo says to no one in particular. "It was comin' out nice and solid, wasn't it?"

He leans against the rig, enjoying the spring weather for a moment.

"Warm in there, huh, John?" he asks as John joins them.

"I knew it was hot when I hit it with the can and I felt it come back to me," John replies.

"The truck got here fast," Ruvolo remarks. "I think they kicked it up a notch when they heard the 10-75."

They go on like this for a while, delighting in every detail of the job. The men all have the same winded look on their faces and the same long dangling web of snot under their noses. That's when the mustaches come in handy. A fire like this, with no one trapped, burned, injured, or dead, is pure fun. It's the kind of physical work that makes up for all the bad stuff they've witnessed lately. And it reminds them why they were drawn to this job in the first place.

# SEVEN:
## FATHER'S DAY

SUNDAY, JUNE 17, 2001, is Father's Day, one of the least popular days to work. It's an assignment that ranks right up there with Christmas and Thanksgiving. (The preferred holidays to work are the Fourth of July and New Year's Eve. You may miss out on the party at home, but at least you get some decent fires.) This particular Father's Day is boring. Nothing is doing, and the guys on duty have spent most of the shift so far calling home to talk to their wives and kids. Each man goes through the ritual of "Congratulations, Dad," then hangs up and sits back down with the guys.

Lieutenant Pete Lund is the boss today. Captain Ruvolo is at home with Janet and the girls. Lund's crew today consists of one of his usual drivers, Billy Eisengrein, and four other solid firefighters: Jimmy Sandas, Jimmy Kiesling, Dave Arciere, and squad 1's Dave Fontana. The guys are lounging around after lunch when the radio emits two tones, then the dispatcher says, *"The borough of Queens is announcing that a second alarm has been transmitted."*

This one's on Astoria Boulevard, close to Manhattan. Traffic is light this afternoon, so it should be a quick drive over there if Rescue 2 is needed. The company doesn't go to other boroughs for ordinary alarms—even big ones; they only go if firemen are trapped or if Queens's own rescue company is tied up at a job. Half the time these excursions don't lead to fire duty; more often than not Rescue 2 gets turned back to Brooklyn on the way. This afternoon another 10-75 is transmitted in Queens. Rescue 4 is already busy at the job on Astoria Boulevard, so Rescue 2 is dispatched to Queens in their place.

Out in Deer Park, Ray Downey is sitting down to lunch at the head of his dining-room table. Downey always takes that place. Seated around the table are his grown children and their spouses. The grandchildren are all over the house, exploring, wrestling, having fun. Downey's two firefighter sons have advanced in the department. Joe is the captain of Squad 18 in Lower Manhattan, and Chuck is a lieutenant in Queens. Even on holidays, the three firemen can't resist gossiping about their jobs. When they get together like this, Rosalie has to shout at them from the kitchen. "No more talking about the job," she yells every few minutes; otherwise they'll never stop. One of the perks of Downey's position as chief of Special Operations Command (SOC) is that he doesn't normally have to work weekends or holidays, unless the department really needs him. These days, Downey only goes to especially large emergencies or fires. He lets the younger guys handle the routine jobs. Nevertheless, he still keeps so many beepers and cell phones around that he could easily be mistaken for a drug kingpin. When something comes over one of the beepers, he will discreetly take a peek, not missing a beat in the conversation.

Today, in the midst of the family gathering, he sees the second alarm in Queens on his FDNY beeper, but then puts the beeper back in its holder, strapped to his belt. The department doesn't need him for a second alarm. But just ten minutes later, the beeper goes off again. Downey glances at it and sees that it's now a fourth alarm with firemen trapped. He gets up from the table, runs into the kitchen to kiss Rosalie good-bye, and mutters to everyone, "Gotta go." Just like that, he's gone. The Downeys have come to expect that he will disappear for hours, days, and sometimes even weeks without more than a "gotta go." If there is a possibility he might be needed, he goes. Most often he gets turned around on the way in to the city because everything turns out okay. As a joke, one of his kids takes Ray's picture off the wall and puts it in his chair; nobody else wants to sit in his place, not on Father's Day.

Downey jumps into his chief's car, switches on the siren, flips on the lights, and speeds into Queens. When he turns on the scanner, he can hear the dispatcher talking to the chiefs on scene. He wonders what he'll face in Astoria.

THE RESCUE 2 GUYS hop on for the long ride to Queens. Pete Lund switches their radio to the Queens frequency so they can hear all the radio traffic from that borough. The job they are going to doesn't sound very exciting, but the other fire on Astoria Boulevard, where Rescue 4 is operating, is getting bigger and better.

The Rescue 2 rig only makes it two blocks from the firehouse before the Astoria Boulevard fire goes very bad. Someone on scene is screaming that Maydays have been transmitted. Someone else gets on and yells that there's been an explosion. Then there are reports of firemen trapped. The dispatcher immediately gets on the radio and reassigns Rescue 2 to Astoria Boulevard. Firemen are down and the dispatcher is sending every rescue company in the city to get them out. Billy floors it. He takes the turns so hard that the side of the rig lifts up off the ground at the apex of the turn. Now every man in the back of the truck is listening intently to the radio as he tries to hang on and put on his gear. Someone turns up the volume so that it is painfully loud. Everything is still vague, but it looks like there is more than one fireman trapped in the burning collapse. The ride to Astoria normally takes about a half hour, even with lights and sirens. But today, after hearing his brothers are in trouble in Queens, Billy makes it there in sixteen minutes flat. The truck has just barely come to a stop when everyone races out the doors toward the enormous plume of smoke.

When they get to the scene, it's total chaos. The hardware store where the blaze started has collapsed and is burning strong. What used to be a hardware store is now just a giant pile of flames and debris. The Rescue 2 guys are the only firemen on this side of the building. A chief motions for them and says that he just saw two firemen get buried after the explosion. Lund leads his men onto the pile of debris. They start to dig through bricks and concrete, searching for firemen. Jimmy Sandas climbs in a window to try and get at the pile from behind. Lund yells, "Come back here. Here he is."

They see a boot and then start to dig even more frantically. The man is completely buried and he must have been like this for at least twenty minutes. He can't be alive. They burrow six feet down into the pile trying to free the firefighter. As they uncover that fireman, they find another fireman pinned underneath him. They frantically try to resuscitate them, but there is no chance of reviving them. It looks like they got hit with the impact of the explosion. Lund recognizes one of the men as Harry Ford, the senior guy in Rescue 4. The other fire-fighter is from Ladder 163; his coat says Downing. Even though the Rescue 2 men have been able to dig around the victims, they are still both pinned underneath a heavy concrete cornice.

"I'm gonna lift the cornice up," says Jimmy Sandas. "Get ready to pull them out."

"You can't lift that thing," someone says.

"Watch me. On the count of three. One. Two. Three." And some-how Jimmy Sandas is able to summon the strength to get that enor-mous piece of building off of these two men.

As soon as they get the two men out of the pile, Lieutenant Lund says, "We gotta go to work." The fire on the other side of the building is burning strong. As the Rescue 2 men circle around and approach the front of the hardware store, they can tell by the amount of fren-zied activity that another fireman must be trapped in there. Lund runs ahead to confer with the chiefs. The other Rescue 2 men join the firefighters already at work, trying to push in and rescue one of their own.

When Lund returns to lead them, Jimmy Sandas asks him, "Lieu, who are we looking for?"

Pete doesn't respond to that question. He just says, "Let's just try and get him out."

But Jimmy notices something in the way Lund answered him. Jimmy won't let it go. "*Who* are we looking for?" he asks again. "Who is it?"

Finally Lund answers him, "It's Brian." Brian Fahey. Jimmy

Sandas's best friend, working today in Rescue 4. Brian is down there. Jimmy has to get him. This is the worst-case scenario for rescue firemen—guys from another rescue company getting trapped themselves. It hits everyone hard, especially Jimmy Sandas.

Jimmy and Brian have been friends since fifth grade, and Jimmy steered Brian to the rescue companies. Brian has only been at Rescue 4 for two years, but, at age forty-seven, with twelve years on the job, he's an experienced fireman. Now Jimmy has to go to work in this fire and struggle to get his friend out alive.

The Rescue 2 men split up and try to find a way to the basement. Jimmy Kiesling and Dave Arciere go to the adjoining building, following Squad 1, to see if they can make it in that way. Kiesling is one of the men Ruvolo has brought over to Rescue 2. The captain worked with him in Squad 1 and knew him as a fireman who could never be stopped, as comfortable hanging off a twenty-story building as scuba diving deep in the ocean. A short guy built like a tank, Kiesling is also an army sniper and seems to know an awful lot of secrets about the U.S. Special Forces, which he never tells. Dave Arciere is a Rescue 2 vet who came out of Ladder 103, the same company as Fred Gallagher, Pete Romeo, and Duane Wood. Dave has a wry sense of humor and often sits quietly on the periphery of a conversation until he chimes in with a devastating remark. In a fire he is entirely dependable and completely fearless. The Rescue 4 guys couldn't wish for tougher men to be coming for them.

Squad 1 has already breached the wall of the basement, and by the time Kiesling and Dave get there, the basement is flooded. When they enter through the breach, the water rises to their waists. The situation is precarious: The surface of the water is sparking intermittently with flames. Flammable chemicals that apparently have been stored down here are burning freely, and it is impossible to predict when the flames will reach chemical deposits mixed in the water. But Kiesling, a glutton for punishment, has the determination of a commando and he is further emboldened by the fact that a rescue man is in danger. He moves forward with extreme care. If anybody can make it through this mess, it's Kiesling.

He seizes the hose line and begins to work his way into the room, with Dave backing him up. When the flames on the water ignite, Kiesling uses the hose to knock them back as Dave, behind him, scopes out every visible inch ahead of them. Brian is almost certainly trapped somewhere ahead, perhaps underwater. Before they can have any hope of reaching him, they must knock down the blaze that crackles ahead. The room is filled with smoke and a powerful cocktail of chemical aromas.

Kiesling and Dave approach the seat of the fire, around a corner to their left. They have to knock this powerful blaze down if they want to go farther into the basement, where they think Brian is. It will be impossible to pass with the fire burning. Attempting to subdue the blaze, Kiesling sweeps the surface of the viscous water with the spray from the hose line and advances toward the corner.

Then, just as he turns the corner, the fire darts out toward him. It skims over the water as quick as lightning and surrounds him. Every fireman knows that it's deadly to have fire behind him. It means there's no way out if things go bad. Dave yells, "It's getting behind you, Jimmy!"

Kiesling turns to hit the flames behind him with the water from the hose. Before the water hits, the fire catches him. He flares up fast because of the chemicals. But he remains calm, as he struggles back to Dave and tosses him the hose. "Knock it down on me, Dave," he orders his friend.

Dave sprays Kiesling full force, and they beat a retreat out of the basement. They know they'll be dead if they continue with this approach. Kiesling is a little toasted, but not seriously injured. His gear has gone brown and he's burned in a few places. But he's not ready to stop trying to rescue Brian.

WHEN RAY DOWNEY ARRIVES, he already knows that the situation is dire. He's heard the radio dispatcher transmitting mixer-off messages, which indicate that a fireman is either badly injured or dead. He jumps out of his Suburban and runs up to the other chiefs to find out the latest. Downey still doesn't know who is trapped. Chief Moran doesn't waste a moment getting to the point: Two men—Harry Ford from Rescue 4 and John Downing from Ladder 163—have already been pulled out dead, and Brian Fahey from Rescue 4 is trapped somewhere in the basement.

Downey can't take in the thought that Harry Ford is dead. A twenty-seven-year member of the FDNY, Harry was the wise old man of Rescue 4, as tough as he was knowledgeable.

But Downey can't think about Harry Ford just now. He has to see if they can get Brian out alive. The chances aren't good. Earlier, Brian transmitted two radio messages. In the first, right after the collapse that trapped him, he reported, *"Rescue irons to Rescue. I'm trapped. I'm in the basement. Come and get me."* Two minutes later, he had spoken again, *"I'm tr— I'm trapped. I'm underneath the stairs. Come and get me."*

It's been half an hour since Brian's last transmission. Worse, during that radio transmission, the other firemen heard Brian's vibra-alert going off in the background, which meant he was running low on air. Odds are against him getting out.

But Downey cannot give up; despite his utter pragmatism, he believes that all things are possible. He is guided on the job by an unshakeable faith. He has lived his life surrounded by death, but he remains a spiritual man who believes in miracles. He was skeptical, at first, when Rosalie asked him if they could go to visit Majagore, a sight believed to have been visited by the Virgin Mary. But something happened to him on that trip. He came back a believer in a way he hadn't been before. After that, he traveled the world with Rosalie to see Fatima, Lourdes, and Knock, all the locations of supposed miracles. Now he hopes for one in Queens, for Brian Fahey.

Wearing that hard, determined expression that his men know so well, Downey surveys the scene. His men are fanned out across the smoking pile, picking up large beams, sifting through bricks and boards. Now he calls them together, as he has done so many times, and firmly outlines their course of action. The men are energized by his presence. This is Ray Downey at his best, calmly taking charge at a time of tremendous danger. He tells his men that they are going to make a systematic six-point attack on the basement where Brian Fahey is trapped: attempting an approach from all four sides, as well as from above and from below.

Downey maps out his plan and assigns men to specific places, giving each a few quick words of advice about what to do. Then they disperse. Though he is sixty-three years old, Downey joins the group heading to the most likely passageway to Brian. His men have been trying, with no success. Now Downey has to try himself. He struggles to make it down a passageway licked by flames, but he has to withdraw, just as his men did. The heat and the smoke are punishing. God help Brian if he is anywhere near this fire.

The guys with the nozzle regroup and decide to make one more desperate push into the fire building, to try and go for Brian. Jimmy Sandas follows behind them. They make it far enough to get into the basement. It is covered in debris. Jimmy and two other men go down the stairs. On his way down, Jimmy sees a helmet and he starts to dig. The helmet says 288, but it must have been knocked off before the guys got out. All the 288 guys are safe outside. Then Jimmy hears a guy from Rescue 3 yelling from down in the basement.

"I got him," he yells up the stairs.

Jimmy walks down the remaining steps and when he makes it down there he and the other two men carefully turn over the body. It's Brian. He knew it was going to be Brian. Ray Downey appears at the top of the stairs. "We got him," Jimmy tells him. Another group of firemen are frantically rushing into the basement, still thinking they might pull Brian out alive. "Calm down," Jimmy says. "There's nothing we can do."

AROUND 5 PM guys start trickling into Rescue 2 for the night shift. When Ruvolo hears what's happening in Queens, he and his men pile into an SOC Suburban and race over to the site. But by the time Ruvolo arrives on the scene, the Father's Day Fire has become a recovery mission, not a rescue.

Ruvolo hears that Harry Ford and another fireman have died, but he doesn't believe it. He thinks it's just a rumor. At big incidents like this, there are always rumors. So and so's dead, so and so's at the hospital. That's what you hear. Yet half the time it turns out that the guys they are talking about are banged up but okay.

Brian Fahey, however, is not going to be okay. From the moment the men discover his body, they fall into ritual. There is nothing to think about; it's like marching, or singing, or saying a prayer. They don't want to think right now. So they follow the routine that firemen in New York have always followed when they lose a man. First they call the members of the downed fireman's company to the body, where everyone removes his helmet and bows his head in prayer. Then the men of the company carry their brother out of the pile.

After everyone is out, Ray Downey gathers all the men and he leads them in a short prayer for those they have lost. Brian Fahey, Harry Ford, and John Downing are all dead. Downey has had no miracles today. Only now, when Ruvolo sees Ray Downey, with his head bowed and the fates of his men written across his face, does Ruvolo really believe that all these men can be gone. What a Father's Day, he thinks, as Downey's voice fades away, coming to the end of the prayer.

Every line-of-duty death hits the guys differently. Ruvolo still has sharp memories of some men who could not be saved—particularly bad deaths, painful or drawn-out scenes linger in his mind for years. The bad ones keep you up at night and make you afraid to meet your wife's gaze or release your child after a hug. *This fire is going to be one of the bad ones*, thinks Ruvolo.

The men will be haunted by the memory of Fahey getting on the radio and calling for help. Trapped in a corner, running out of air and watching in terror as the flames advanced toward him, he had called out, *"Come and get me."* He had suffered, suffocated, and burned. But the men know better than to let that image get in their heads; it's too much to bear.

And it is Father's Day. Somehow that just seems too cruel. Standing on the sidewalk next to the smoldering pile, Jimmy knows that he will now have to take responsibility for Brian's family for the rest of his life. It will be an honor, but a sad honor, to go to soccer games, teach the boys how to throw a curveball, and do all the stuff Brian won't be able to do. This is how it works in the FDNY. It has always been this way; the men's commitment to their brothers goes further than death. It must be that way if they are to continue in this line of work. Each man must know that if he dies doing what he loves, his family will be taken care of.

Losing Harry is a different kind of tragedy; not the particular kind of loss that comes with a young man's farewell, but no less difficult, because of all the years to remember with him. It's a bitter thing to say good-bye to the toughest, most experienced guy in the house. He keeps popping up in stories; you keep wanting to ask him how to do something, and he's not there to ask.

On top of all this, it's Father's Day. Every one of these men should have been getting home from their day tour right now and sitting down to dinner with their families.

Every fatal fire is given a name that the men use when they refer to it. A name gives a fire its place in the history and lore of the FDNY. A name might memorialize a fireman who died fighting the blaze, or it might refer to the fire scene. The fire that killed Louis Valentino is always called "The Louis Valentino Fire." The fire on Atlantic Avenue is always called "Atlantic Avenue." But the fire that killed Harry Ford, Brian Fahey, and John Downing will always be known as the "Father's Day Fire," a name borrowed from the headlines of the *New York Post* and the *Daily News*.

The Father's Day Fire hits the men of Rescue 2 with severe force. Despite their long-standing rivalry with Rescue 4, the other rescue companies are still their closest comrades on the job. So the day before the funerals, Phil Ruvolo calls over to Rescue 4 and offers to have the Rescue 2 men cover at Rescue 4, so all of the Rescue 4 men can attend the funerals. It's something the rescue companies do for each other when a line-of-duty death occurs. The Rescue 2 men don't get paid to do this; they do it out of respect for Brian, Harry, and the firemen they've left behind.

Outside of Brian Fahey's funeral, Jimmy Sandas's wife comes up to Chief Downey.

"Do you know what they call you?" she asks him. "They call you God."

"There was one day I couldn't be God," Downey tells her.

After the Father's Day Fire, where he lost his best friend in the world, Rescue 2's Jimmy Sandas calls up Ray Downey every day of the week. At first he calls about administrative details: funeral arrangements, paper work, widow's benefits. But gradually the conversations move to a different territory: the meaning of Brian's death, how Jimmy feels after losing his best friend, and what he can do to help raise Brian's boys. At first Jimmy feels bad about bothering Downey. But after a few conversations, he realizes that the chief needs to talk too. Downey didn't know Brian like Jimmy did, of course, but he feels this loss of one of his own men acutely. It helps Downey to talk about it with someone who understands his pain, and has pain of their own.

AFTER COMING HOME from two of the funerals, Kevin O'Rourke goes with his wife to pick up their younger daughter Jamie from a party on the day after her prom. Kevin and Maryann watch as their daughter dances with her boyfriend; the next dance is Kevin's. Maryann watches as her husband leads her daughter across the dance floor. As he comes toward her Maryann can see his face, and suddenly she realizes that Kevin is crying. Maryann interrupts and takes him aside. As Kevin heads outside to get some fresh air, Jamie turns to her mom and asks, "What's wrong with Daddy?"

"He's just sad to see his little girl all grown up," says Maryann. But she knows it isn't just that. The Father's Day Fire hasn't just upset Kevin; it's destroyed him. Ordinarily, rescue firemen feel invincible. But when a rescue man dies a horrible death, nobody feels invincible anymore. Every guy realizes: there but for the grace of God go I. And this time Kevin has to realize all that on Father's Day, and think even more about what his death would mean to his children.

The thing that starts to turn things around for Kevin is a fluke. The same summer of 2001, Kevin O'Rourke is sent to Manhattan to work overtime on a day tour at Rescue 1. But first he has to stop by

Rescue 2 to grab his gear. He packs his gear into the car and journeys to the city. When a fireman goes on a detail, he always leaves his firehouse by 9 AM, arriving at the company he is detailed to after the shift has already started. This gives someone else a little overtime.

Kevin makes it to Rescue 1 at about 9:30 in the morning, parks his car outside the firehouse on West Forty-third Street, and trudges up to the front door, which is already open. The guys are, at that very moment, hopping onto the rig for a run. Kevin darts inside the firehouse and jumps on the back of the truck, just before it pulls out. The Rescue 1 guys don't even see him get on. He hears the radio crackle with details of the run. Someone is drowning in the Hudson River, fairly close to shore.

Kevin opens up a compartment in the back and finds an exposure suit to put on, so he can float in the water. Exposure suits are used for surface rescues, where the victim hasn't gone underwater. The Rescue 1 guys still don't really notice Kevin. Wriggling into the gear, he's way out of sight in the back of the rig. That's okay by Kevin, who has a plan in mind, a scheme to announce that a Rescue 2 man is on the job. If he manages to get into the water before them, he will score a surprise victory for Brooklyn. Competition between these boroughs is beyond fierce: Manhattan protects the giant skyscrapers of the city and looks down on the little buildings across the water, where Rescue 2 does its work. The men of Rescue 2 see the Manhattan guys as prima donnas who don't see much fire action and work in a borough that considers itself the center of the universe.

When the rig pulls up to the river, Kevin spots the drowning guy immediately, struggling near some half-submerged pylons. The other guys pile off the rig and start to assemble a rope system to lower a rescuer down. Kevin, however, doesn't wait. When he hears somebody call for a guy to go in the water, he climbs over the fence and jumps. He hits the pylons hard and bangs up his arms and legs. Still, he keeps swimming. When he reaches the drowning man, he drags him to shore, where the other guys are waiting to hoist him up. It's not until this moment that some of the Rescue 1 guys realize there is an intruder in their midst. They hadn't realized that Kevin was on the truck. Now he's stolen one of their grabs. But this petty rivalry is just that; and all the men are happy they got the victim out alive, no matter who made the rescue.

After losing three firemen, it makes Kevin feel especially good to save a life. These are the small compensations all the guys collect to try to stem the tide of loss.

AT HOME IN RONKONKOMA, John Napolitano just rocks back and forth in the chair in his bedroom. This is how it has been since he returned from the funerals. Ann doesn't know what to say to him and she understands that her husband doesn't want to worry her by talking about what happened. He is so upset and uncomfortable that it's painful to look at him. To Ann, he looks like he's on fire inside.

At Rescue 2, John Napolitano finally thinks of himself as one of the guys. When he starts talking to Ann about getting a Rescue 2 bulldog tattooed on his arm, she realizes he's made it: now he's truly a part of the rescue brotherhood.

John, who is a good artist, who spends days sitting on his living-room floor, painting a stained-glass Rescue 2 sign to put in Phil Ruvolo's office door. He has also rebutted Larry Gray's cartoon offensive against him with a cartoon of his own about the old man. Most recently, he has set about drawing a Rescue 2 bulldog with which he intends to decorate his arm for the rest of his life. Early in the summer, after the Father's Day Fire, he visits a tattoo place in Suffolk and goes through the whole bloody ritual. When Ann comes home, he surprises her with the tattoo. First he lures her into the kitchen; then he exposes his arm. He keeps the video camera rolling all the time to record her reaction. She's shocked, but not exactly happy, only after a long pause does she manage to squeeze out a smile for the camera.

John works hard all summer because the tenant of the downstairs apartment in their house has left, and they've decided not to find a replacement. So John has to make up the lost rent by doing even more work on the side. It seems he's done almost everything to support his family: driven a limo, bartended, worked construction. This summer he works three jobs—FDNY fireman, bouncer at the Boardy Barn, and painter, for a friend's company. In his few free moments he also studies for the upcoming lieutenant's test.

The days when John first came to Rescue 2 now seem like a bad memory—ancient history. Back then, Ann was pregnant with their second child, and he didn't want to even talk about work for fear of upsetting her. Now that child is a little girl of two and a half. When he's at the firehouse overnight, John still calls home twice—first, early, to say good night to his two daughters; then later to say good night to Ann. Since the Father's Day Fire, he too is especially aware of how precious his family is, and he wants them to feel secure.

That summer, John finds a dog abandoned in front of the firehouse. He takes it in and gives it some food and water. Then he calls his dad to talk about the dog. Finally, he says, "Do you think Ann will mind if I bring the dog home?"

"If she does, I'll take it for you, Johnny," his father tells him.

So John brings a Brooklyn mutt home to Long Island to guard his girls while he's away at work.

LIKE EVERYONE ELSE, Ruvolo is still getting over the Father's Day Fire. It helps immeasurably that July turns out to be a good month for him. Although much of what he has experienced with his men has been very difficult, the events that have made his time as captain so hard have also united him with the house. He feels more relaxed about being in charge. His spirits start to lift in his old home turf of Staten Island, where he picks up an overtime shift at Rescue 5 one muggy summer afternoon. Despite his captain's status, he needs extra income and, like all firefighters, picks up overtime whenever he can. Overtime pays time and a half for what he loves to do and it's much better than banging nails. Overtime at Rescue 5 is a lucky break for Phil Ruvolo; it's just a ten-minute ride from his house and he knows the men in the firehouse well. He did time here as a lieutenant back when his father-in-law was dying and he needed to be close to home. Many of the Rescue 5 guys took that posting because it put them just a few minutes away from home. Compared to commuting for an hour or two, it's paradise. The only drawback is that Staten Island can be slow, even for a rescue company, and the large size of the borough, and its notoriously bad traffic jams, can mean long delays getting to fires.

But on this afternoon, the dispatcher comes over the radio with a report of a flurry of phone calls about a fire on Townsend Avenue, near the Staten Island Ferry Terminal. Ruvolo slides down the pole and climbs into the cab.

This box on Townsend Avenue sounds promising. Rescue 5 makes it in third, after an engine and a truck. They work as a second-due truck company, which means Ruvolo is assigned to the floor above the fire. He leaves his can man downstairs with the engine, to help those guys and to protect him since he's going above the fire, the most dangerous position. The building is a large private dwelling: a big old house that's been converted into apartments. Ruvolo enters the front door, behind the engine company. The engine doesn't have water yet, but they're getting ready for it.

Inside the house, the fire is burning through one apartment door into the hallway. It's a narrow hallway, and Ruvolo needs to pass the fire to get to the stairs, so he ducks down and hugs the right-hand wall, away from the fire. It's warm going by. As he hits the stairs, it gets hot and shitty, smoke all over and heat banking down from above. On his way up the stairs he takes off his helmet with one hand and slides his mask over his head with the other. He's gonna need it up here.

Often Ruvolo will have a forcible entry team with him to take the door. But this time his guys are busy downstairs, so he's on his own as he attempts to enter the apartment directly above the fire. He wrenches his Halligan tool's claw into the doorjamb next to the lock and then pulls it toward him, trying to pop the lock. Then he slams it downward for good measure and takes the door. Inside, he's on his hands and knees in a crawling sprint to hit every inch of this apartment before the smoke and fire kill everyone inside. The fire from the floor below has spread up the old wood walls, and it's going good in this apartment too. The walls and ceilings are burning, and the place is thick with black smoke. If any victims are in here, they're out cold.

Ruvolo crawls down a hall to make his search, and after fifteen feet he hits something on the floor, in the doorway to a bedroom. It feels like a body, but you never know until you check; sometimes a duffle bag will feel like a body. But when Ruvolo feels around with his gloves, he knows it's a victim. She must have passed out trying to escape.

Ruvolo radios in, *"Rescue Officer to Battalion, 10-45, I've got someone upstairs, Chief."*

Then he turns around and starts to drag the victim, any way he can. A few feet from the upstairs apartment door he runs into two other guys from Rescue 5 and passes them the victim. They'll take her down to the street to EMS, to try to revive her; Ruvolo wants to go back inside and search for more victims. "I'm going back in to see if there are any more 10-45s," he tells the guys.

The apartment is lighting up, and it won't be long until the whole place is whipping with flames. Right now the walls are red, the ceiling is roaring with fire, and the attic above the ceiling is shooting out flames. But Ruvolo works methodically, inch by inch, staying close to a wall to orient himself while he sweeps the whole place with his arms and legs. By the time he finishes, he can hear the engine company coming in behind him. Ruvolo can help the guys knock this thing down by directing them to the fire. The rule is you don't open the nozzle until you can see fire. But because the thick black smoke is obscuring their view, they will have to get close to the fire to see it. "Come on down," Ruvolo yells. "Fifteen feet down, on your right."

He's directing them to the seat of the fire, where it's burning strongest. If they start there, they'll put it out faster. Ruvolo is stuck in front of this engine company, and he hopes they don't open the hose line until they've gotten past him. If they hit him with the water, he'll be pissed. A strong stream will knock off his helmet and throw him around. That's called getting power-washed, and if an engine team is pissed at a truckie, they just might give him a power wash.

Ruvolo doesn't get power-washed this time. The engine moves in and starts to knock the blaze down. After the engine hits the main body of fire, the truckies move in and pull the ceilings, exposing any fire hiding up there. Pulling ceilings can sometimes be used as payback: if a truckie gets power-washed, his easiest revenge is to pull ceilings over the engine company's heads, sending burning debris on top of them. It might get him power-washed next time, or it might serve as a deterrent. But today everybody plays nice. The fire goes just as planned. They knock it down, chase it up the ceiling, finish it off.

Ruvolo heads downstairs to inquire about his victim. He learns that the paramedics were able to revive her in the street. He also gets the name of the hospital where they took her because the Fire Department doesn't notify rescuers of the fate of the fire victims they pull out alive.

Later that night Ruvolo calls up the hospital himself, to check on the woman he pulled out. "You had a woman, a fire victim brought in today. I'm the fireman that found her," he says. "How's she doin'?"

"She's critical."

"All right. What's her name?"

He'll never call her directly, and he'll always just speak to the operator; but he will call every day for the first week, just to see how she's doing. Making a grab is exhilarating, but right now, after dragging her out, Ruvolo is tired. He's lucky. It's hard to make a grab. Many guys go their whole career without pulling anybody out; some guys have a knack for finding people. They always seem to turn right when everybody else goes left, and find somebody there. Patty Brown has that knack. Other guys go to the same fires, do the same searches, with the same intensity, but come up empty-handed. If there's a body in there, it's Patty Brown who finds it, more often than not.

Ruvolo doesn't have Patty's luck, but he's found a good few bodies over the years. Finding a person alive today starts to make up just a little bit for the pain of the people he wasn't able to drag out still breathing, like Brian Fahey.

TEN DAYS LATER, Ruvolo is alseep on his bed at Rescue 2 (he's upgraded Downey's flimsy old cot) when he's awakened by an alarm. It's a first-due box on their street, about five blocks down, toward the city. He's on the rig in thirty seconds flat. There aren't multiple telephone calls coming in to report the fire, so it's hard to tell if it's really a fire or just a false alarm. But since the fire is supposed to be in the rear of an apartment building, it might be back where hardly anyone can see it. Rescue 2 arrives on the scene first, which all the guys love. This is the most exhilarating moment of the job, when there's nobody around but you to save the day.

Bob parks the fire truck a few houses away from the apartment building, leaving room for the engine and truck to get close. From the outside, they can't tell if there's a fire, so Bob will stay with the rig until the other men determine whether the building's on fire. Ruvolo, John, and Kevin go inside. The roof man heads for the roof. And the OVM heads up the fire escape, breaking windows as necessary.

Luckily, the front door to the street is open, so Ruvolo and his two colleagues don't have to hit buzzers to try and get a sleepy resident to let them in. They sprint up four flights of stairs. As soon as they get to the hallway, they can tell they've got a job. Black smoke is pumping out of the rear apartment door, around the frame, below the door, and through every crack. Just from the way the smoke is pumping, he can tell this is more than food on the stove or a mattress burning.

"*Bob, transmit the* 10-75," Ruvolo radios to Bob, who's got a borough-wide radio down at the rig.

Bob sends out the 10-75: a real structural fire. In this situation, Rescue 2 is working as the first-due truck company, so their assignment is to search and vent the fire apartment.

Ruvolo puts on his mask. Kevin and John get ready to pop the apartment door. Kevin has the ax and the Halligan with him, and John has the fire extinguisher, which might buy them some time while they search. Kevin hands the ax to John and positions his Halligan in the door so that John can strike it with the blunt side of the ax and smash open the lock. They take the door and dash in for a search. John and Kevin go straight and quickly disappear into the smoke. Ruvolo follows.

Ruvolo can see about ten inches in front of his face. He catches a glimpse of a hallway off to the right. He doesn't know if it will take him anywhere, but he figures he'll try that way, since the others went straight. Ten feet down the hallway to the right he comes to a door. When he tries to open the door, it feels as if something heavy is behind it, maybe a body. He shoves harder and gets in. It's not a body, just bags of clothing piled up behind the door. He starts around the room and then hears a television set and thinks, *Someone must be in here.*

He sweeps the floor, then reaches up on top of the bed to feel for a body. Nothing. Only one place is left in the room: the other side of the bed, next to the wall. He goes over to the bed and checks there. Bingo. A body. A big one. There's no way for Ruvolo to get this victim out except to drag her over the bed. He picks her up and lifts her onto the bed, then yanks her off the other side. Now he starts tossing bags of clothing away from the door, so he can get her through.

Ruvolo can see the fire through an open doorway into the rest of the apartment, a different one than the one he came in through. The fire is about to reach into this room, so he needs to get her out fast. He drags the body out the door and starts down the hallway. While every fireman's dream is to grab a dainty little maiden who weighs 100 pounds; somehow it never works out that way. This woman must be 250. It's backbreaking work to move her. Ruvolo crawls to the apartment door, with the body in tow, then stands up, grabs her around the waist, and starts to exit. But he trips on the door saddle and tumbles out into the hallway, with his 250-pound victim falling on top of him.

The first thing he hears out in the hallway is someone from the engine saying, "I've never seen that before." He looks up and sees a kid with a probie's shield.

"Don't just stand there," says Ruvolo. "Get her off me."

Ruvolo and two of the engine men carry her down four flights of stairs, out to the paramedics. The EMTs bring her back to life and speed away to the hospital.

It's not written down in any department regulations, they don't teach it to you in probie school, it's not mentioned in the newspapers, but Ruvolo and the guys like him have a rule about rescues. It only counts if the victim survives for twenty-four hours after the rescue. If the victim dies in the street, it was just another day's work. If the victim makes it twelve hours but dies in the hospital, no good. But if you call up the next day, after twenty-four hours have passed, and the nurse tells you, "She's still alive, in the ICU," you get your grab. That means you can talk about it, and you can put in for a meritorious act, a department designation for when you save someone. Ruvolo is on the verge of two rescues in ten days. On the verge, because he doesn't know if this one will last twenty-four hours.

A week later Ruvolo, John, and Kevin are hanging out in front of the firehouse. It's a hot summer day and they've got the front door open to let in some air. A black man in his late thirties approaches them and says, "Some firemen saved my mom about a week ago, down on Bergen."

"That was us," and Kevin points to Ruvolo, John, and himself.

"I don't have a lot of money," the man says. "But I have two thousand dollars, and I'd like to give it to you guys for saving my mom."

"We don't want your money," replies Kevin. "But you know what you can do with it? You can buy everybody in your family a smoke detector."

"Okay," says the man, a little taken aback that the firemen don't want his money. "I will."

"Oh wow, that's a great deal, Kev. The guy thought he was going to part with two thousand bucks, but now all he needs to do is buy a couple of smoke alarms. Not bad," jokes Ruvolo, after the man is gone.

Somebody else would have just said, "Glad everything worked out. Don't worry about it. We don't take money like that." But Kevin tells the guy to buy smoke detectors with the money. It figures.

Now Ruvolo has two calls a night to make to hospitals. His first victim is still in critical condition, but this second woman seems to be doing better. It sounds like she'll be out of the intensive care unit in a few days.

A week later the Brooklyn fire victim leaves the intensive care unit. Ruvolo still calls up the burn center about the Staten Island victim. She'd been there, in critical condition, for over a week now. He's hoping she'll live. And he'll keep calling every week without ever saying a word to her. All in all, Ruvolo's starting to feel a lot better. Things are picking up.

AFTER THE FATHER'S DAY FIRE, Downey calls up Al Fuentes, just as he did after the day he saw a probie burn to death in an upstairs window. He asks Al to meet him on Astoria Boulevard, at the site of the Father's Day Fire. Downey is still badly shaken. He hides it from the men, but Al can see that Downey is having a hard time coming back from this one. Al listens carefully as Downey walks him through the fire, points out where Brian was trapped, explains how the firefighters had tried to get to him, and recalls where each rescue and squad company was operating. It seems to Al that Downey is not merely trying to teach him something; he also needs to go through it all for himself. He has to figure out if there was anything he could have done just a little bit differently that would have saved Brian Fahey.

The deaths hit Ray Downey hard. This is the first time Downey has ever buried two of his men. One was bad enough; it had taken years to recover from Louis Valentino's death. For a veteran like Harry to die, so close to his retirement, never having done all those things he was always meaning to do after he retired—that hurt. And to know that Brian Fahey was down in the basement, waiting for them to come for him—that made it even worse. In the days that follow, Rosalie looks out her window to the backyard and expects to see her husband cleaning up the pool or mowing the lawn. Instead, she finds him just sitting and thinking. The Father's Day Fire has affected him more than any other fire she can remember in his thirty-nine years in the department.

One day Rosalie finds her husband sitting on the couch, crying. "What is it, Ray?" she asks.

"I lost my men," he tells her. "And I'll never be the same."

Chief Downey has a lot on his mind. Fires, even bad ones like this one, are just the beginning of what he worries about. Worrying is part of his job description. If he doesn't worry about what might happen to his men and the citizens of his city, he's not doing his job. He had been leading the rescues at the first World Trade Center attack in 1993, right after the bomb exploded. He had been at Oklahoma City in 1995, in charge of the FEMA operation, just hours after that attack. And he had been on scene in Atlanta in 1998 for the Olympic bombing, again working for a FEMA team. On his desk are all the reports he can get his hands on about the threat of terrorism. Downey has become convinced that a major terrorist attack is coming, and that very few people in New York, or the United States, are prepared for this eventuality. So, on top of worrying about more of his men going down in fires or collapses, he worries about terrorism too. He can't do much to prevent an attack, so he concentrates on preparing a team of men who will be able to protect the civilians and the firefighters who might be hurt in an attack.

I N 1995 Downey left for Oklahoma City, much as he had for the Father's Day Fire; he glanced at a beeper, said "Gotta go," and later called Rosalie from the airport to tell her he would be away for a while. In fact, Downey wasn't surprised; he'd expected something like this and actually always kept with him a bag packed with underwear, extra clothes, and a toothbrush. Hours after he was sent out there to run the rescue operation, he activated his own New York Urban Search and Rescue team to fly out and assist in the rescue efforts, bringing many of the guys he knew from his days as captain of Rescue 2: Richie Evers, Billy Lake, Jimmy Ellison, Al Fuentes, Pete Martin, and Dennis Mojica.

The toughest thing about Oklahoma City was that the FEMA teams didn't find anyone alive. In the first days there was adrenaline flowing, exhilaration at the thought that if they could get in there fast, they would pull people out alive. But as the days passed, Downey and everyone else realized that this was a recovery mission. Then it became something very different from rescue work. There would never be a moment of ecstasy, when a searcher found someone alive. But still, the search team saw it as their sacred duty to the victims' families to recover the bodies of their loved ones. When Downey's men uncovered a pair of pants with a marine's red stripe on them, Downey stopped the operation until they could get a marine color guard to assemble and salute as they took out the body of this fallen marine. More than thirty years after he'd gone to Parris Island, Downey still felt like a marine.

There were many technical challenges at Oklahoma City—like

how to get by the "Mother," a giant concrete slab that loomed over the rescuers, threatening to cascade down at any moment. Downey overcame the technical challenges and found a way to command a huge team of men from all over the nation. He found that he liked leading this kind of large-scale operation. At a big job like this he got to use all the skills he had acquired in decades of fires, collapses, and emergencies in Brooklyn. No firefighters anywhere in the country were as used to urban disasters as Downey and his men. He was proud of the way the men worked, fast and hard, but always protecting the dignity of the dead.

One day, Downey was standing in a makeshift office near the Oklahoma City site talking to FEMA head James Lee Witt when Bill Clinton came into the room. The president took one look at James Lee's cowboy boots and said to Downey, "What do you think about this? James Lee doesn't have a pair of shoes." Clinton and Witt were old pals from Arkansas. Downey shook hands and exchanged a few words with the president. When Clinton came down to the site, he sought out Downey again. Downey had a confidence that politicians found reassuring.

Downey and Al Fuentes, who had shared so many things, were both blown away by the kindness of ordinary people in Oklahoma. When they walked away from the site in the early morning, after a whole night of work, there would always be a volunteer around who would ask, "You want a hug?"

At first Al thought, *What? A hug?* But when he got one it felt good.

It became a running joke between the two men: Downey would turn to Fuentes in the middle of the operation and say, "You want a hug?"

After the New York team had been there for a week, Downey got the order to send them back home and rotate in another team. He called aside Al Fuentes and told him, "Al, you're leaving. Tell 'em one more hour."

The New York team was digging in one area and was close to the bottom. When Al told the guys they soon had to go, they flat-out refused. "No," someone told him. "We're not leaving until we get to the bottom of this hole here."

Al turned to Downey and said, "The guys don't want to leave until they get to the bottom of the pile."

Downey couldn't decide if he was pissed off or proud. Even here in Oklahoma, commanding the whole operation, he still had to put up with a bunch of stubborn bastards from Rescue 2. But he understood why the guys were defying his orders. He had taught them well. "You fucking tell 'em they got two hours to get to the bottom of the pile," he told Fuentes.

"Guys, the man said you have two hours," Al reported. He saw a little smile start to creep across Downey's face. Al knew that his friend was secretly proud of his men. When they were about to leave, Downey got his men together and they prayed. It was something he always did, but it seemed especially important here, in the midst of the wreckage of such an evil deed. The whole thing was horrible, but one image in particular would haunt Downey long after Oklahoma City : the bodies of little children being recovered from the wreckage. They were innocent children who went to a day-care center at the Murrah Federal Building. Downey had seen several grown men die trying to save the lives of others. But as hard as those deaths were to accept, they were sacrifices made for a higher purpose. The children in Oklahoma City died for no reason at all.

There and then he made up his mind: if something like this ever happened again, he would be as ready as he could be. Because the more prepared you are at the moment of impact, the better chance you have to pull people out alive.

In Oklahoma, he felt the world changing under his feet. Clearly this could happen again on American soil and kill a lot of people. And if there was a prime target for terrorists, it was Downey's hometown. People had been able to write off the 1993 World Trade Center bombing as an unsuccessful attack, a demonstration of the colossal ineptitude of the terrorists, who watched from Jersey, waiting for the towers to come down, but only killed six people. But two years later, Oklahoma City proved that it wasn't hard to kill a lot of Americans;

even an unsophisticated fertilizer bomb could kill 168 people. Who knows how many a smart terrorist could kill?

AFTER OKLAHOMA CITY, Downey got the job he was looking for in New York: commander of the newly formed Special Operations Command. In addition to his obvious task of planning the FDNY's response to terrorist attacks, Downey had some very specific concerns. He'd commanded men at the World Trade Center back in 1993 and he'd seen how dangerous the job was for the rescuers who responded. A firefighter from Rescue 1 had fallen thirty feet into the bomb crater and had to be rescued himself. Downey was determined to figure out ways to protect the rescuers and make sure they didn't get taken down too.

In 1998, when the Tokyo subways were hit with poisonous Sarin gas, Downey considered what his men would do if a similar attack occurred in New York. After the Tokyo attacks, when Mayor Giuliani had asked Downey, "Are we prepared?" Downey had to answer, "No, we're not."

But immediately he got working on a plan for subway emergencies. One more thing to worry about. The job description was growing. Downey quickly realized that dealing with nerve gas was a very different job than knocking down a blaze or pulling off a rope rescue. A few months later, Downey met with Fire Commissioner Tom Von Essen. They worked out a plan to convert six engine companies into squads that would be under Downey's command, just as the rescues were. The new squads would carry some rescue equipment, but their most important emergency function would be to deal with hazardous materials. In the event of a chemical attack, they would be equipped with protective suits and decontamination units. Each squad would house an additional "haz-mat" truck that could be driven to an emergency. Downey knew that this kind of unit was badly needed by the FDNY. If an ordinary engine or truck responded to a chemical attack now, the first wave of firemen would run right in and die. There was already one haz-mat company that did this kind of work exclusively, but that single company was serving a city of eight million. The six new squads would spread this capacity throughout the city.

Downey worked out various scenarios for terrorist attacks—who would be the first, second, and third of his companies on scene; what would each unit do. He studied floor plans of major landmarks, looked at aerial views of the city, thought about traffic routes, bridges, and tunnels. He shared this work with some of the men he knew best. He had moved men he trusted into key positions throughout the city, and if an attack came, he knew he would be surrounded by them. Downey sent one member of his inner circle to work at the Office of Emergency Management, the group that coordinates the city's response to disasters. Al Fuentes became the captain of the Marine Division, the fire boats. When Downey and Al started to get information about threats against ferries in New York, they scrambled to prepare for that eventuality, conducting drills for a chemical or dirty bomb attack on the Staten Island ferry.

Downey, who had witnessed so much of the history of the rescue companies, realized that the companies were approaching a level of sophistication and experience that had never been seen in the city. Downey and his generation had weathered the war years and used all that first-hand experience to create unstoppable rescue companies. Now it was time to make certain that the wisdom of their special generation of firemen would not be lost. The future might not provide younger men with the experience that had so benefited the older generation. Just as the U.S. military, in the time between wars, focused its efforts on intensive training, Downey wanted to do the same for his rescue companies, so that, when they were needed, they would be prepared.

At first he got a little shack at the Rock, the FDNY's training ground at Randall's Island. And then they began that slow process of institutional scavenging they all knew so well from years of living under a frugal city bureaucracy. Downey scrounged up some desks, books, and equipment. They started to write a curriculum and teach classes. A rescue school took shape.

Downey was still interested in fighting fires, of course, and jobs like the Father's Day Fire reminded him just how important that work was. But after witnessing the aftermath of a rash of terrorist incidents on American soil, Downey felt certain that a big one was coming next. The first World Trade Center bombing had demonstrated that Islamic terrorists saw New York as their prime target. To Downey, the question wasn't if another attack was coming, it was when and where. "We're gonna get hit bad," he warned, whenever the subject turned to terrorism. He had discussed it with all his men and run down the scenarios. Ray Downey and his men thought that the big one would most likely be a chemical or dirty bomb attack in an urban environment. Worst in their mind was a situation in which there was a secondary device timed to explode after rescuers had rushed to the scene.

These were gruesome things to think about, and it required a particular mentality to want to be the guy running toward a dirty bomb or a subway gassing when the whole city would be fleeing in the opposite direction. But Downey knew where to find that mentality—the guy who thinks he's invincible, the guy who knows the special adrenaline of the grab. He could find it in the rescue companies—and they would become the crack teams Ray Downey needed to keep his city safe.

WHEN DOWNEY TRIED to tackle the problems of responding to a terrorist attack, he liked to have the guys he knew best working at his side. As captain of Rescue 2, Downey had been tough on the men under his command. Even the firemen who liked him best were a little bit afraid of him. But after they got promoted, this changed. Whenever one of Downey's Rescue 2 firemen got promoted to lieutenant, he would notice a difference in the way Downey acted toward him. Now Downey was helping him out, finding him a spot, and putting him to use in SOC because men whom Downey trusted were his most important asset in building SOC into what he wanted it to be.

In this intense effort to guard the city, Downey turned to Phil Ruvolo and Terry Hatton for help in building the curriculum of the newly formed Rescue School. Hatton was the son of an FDNY deputy chief, a Long Islander who had dreamed of joining the Fire Department from the time he was a little kid and who approached his job as studiously as if it were a tenure-track faculty appointment. Hatton read and studied everything he could about rescue work. When he first came to Rescue 2, Hatton dove right into the tools and mastered most of them in a couple of months. He was a tall man with a baby face but a commanding voice and when he became an officer, he quickly earned a reputation similar to Downey's: that of a relentless taskmaster who was fair—but firm—with his men.

A decade earlier, Terry Hatton had been the young-looking kid at Rescue 2. Now, in 2001, he was the young-looking captain of Manhattan's Rescue 1. Yet, despite his appearance, Hatton was mature, and, thought Ray Downey, the man best able to lead Rescue 1.

Hatton had recently married Mayor Giuliani's assistant, Beth Petrone. But even when Hatton took his wife out on the town in New York, he kept an eye on his beeper. If he saw a good fire or emergency come in, he couldn't resist. He would apologize to Beth, then drag her across the island of Manhattan to catch a glimpse of a good job in progress. Hatton didn't go to these jobs to lend a hand; he just wanted to watch the other guys at work and figure out how to do his job better.

Hatton also was the most anal-retentive fire officer Downey had ever seen; coming from Downey that was quite a distinction. Everything he read was filed, organized, and taken care of immediately. But this was part of what made him great. He was on top of everything. He read every magazine, cataloged every job he ever had, went to conferences, spoke to manufacturers, soldiers, politicians. And he did this all with a single-minded purpose: protecting the city. Hatton was a man Downey was proud to command and one he could count on in a crisis. For to be the captain of Manhattan's Rescue 1 in the twenty-first century was a very serious thing: when the "big one" struck New York, odds are it would hit Manhattan.

IN JULY 2001, Ray Downey, his family, and his closest friends on the job gather at Gracie Mansion. Tom Von Essen and Rudy Giuliani had decided to honor Ray Downey with the Golden Apple Award, which the mayor gave to New Yorkers he loved. Downey is dressed in civilian clothes—a tan summer suit—and he is beaming with pride. Many of Downey's old buddies are there, including his chauffeur, John Barbagallo, who is awed by the whole thing and couldn't be any happier if he was being honored himself. At one point Downey looks up and sees Ruvolo talking with Hatton. There they are, two men of his inner circle: the captain of Rescue 2 speaking earnestly to the captain of Rescue 1. Downey can't wipe the smile off his face tonight. To be surrounded by all his children and his wife, Rosalie, on an occasion like this is pure bliss, the payoff for a lifetime of hard work. Who would have guessed that a tough kid from Woodside would one day be wined and dined by the mayor of New York?

# EIGHT:

## THE TOWERS

O N THE MORNING OF SEPTEMBER 11, 2001, Bob is taking a test at the city's EMS Academy in Fort Totten, Queens, just under the Throg's Neck Bridge. Today he is the only person taking the two-hundred-question, multiple-choice exam for recertification as an emergency medical technician. One proctor monitors him, occasionally stealing glances at him when she looks up from her newspaper.

"Do you wanna take a break?" she asks. "The coffee truck is coming."

"Yeah," Bob replies. "That's a good idea. I could go for a cup of coffee. Just let me get to question one hundred."

Soon after, they walk out to the parking lot, where thirty identically clad EMS workers are lined up around a shiny metallic concession truck, with a coffee urn pouring from the rear, cold drinks on one side, and hot breakfast sandwiches on the other. Bob's pager starts to beep, and, as he fishes it out of his pocket to see what's up, almost all the paramedics do the same. (Many of the city's firemen and paramedics carry "buff pagers," which alert them to all the big fires in the city.)

"Oh, wow, a plane crashed into the trade center," Bob hears, as he sees the report on his pager. *Must be a Cessna, something small,* he thinks to himself.

Two minutes later, as they stand around discussing the situation, the chief of the EMS Academy comes out and announces, "All right, everybody, listen up. We just had a commercial plane crash into the trade center. We're loading up ambulances, and we're going downtown."

*Commercial plane?* Bob thinks. *I gotta get down there.*

"Rescue," the chief says to Bob, "I'm going down there now. You wanna come with me?"

"Yeah, sure I do," says Bob. If they leave now, in an emergency vehicle, they'll be there in ten or fifteen minutes. Bob will meet up with the Rescue 2 guys who are responding, and fight this fire with them. Sure he wants to help out, but he also doesn't want to miss this one. It could be huge, legendary: the kind of job that will be kicked around the kitchen for the rest of Bob's career.

But when Bob follows the chief into an office, he gets a look at the trade center on television. Flames engulf whole stories of the North Tower, smoke is pouring out, papers are flying everywhere in the bright blue sky. After one look at that, Bob turns to the chief and says, "Listen, thanks for the offer, but I gotta go back to my firehouse for my gear. There's no way I can go to this thing in a pair of dungarees and a T-shirt."

"You sure?" asks the chief. "I'm going right now."

Bob hesitates. Maybe he could pick up some gear off of one of the rigs. No, he should go to the firehouse. "I gotta go to my quarters," he tells the EMS chief. "I'll see you down there."

It's a big job, maybe the biggest of his life. Bob does ninety across most of Queens. He doesn't want to get there late. He can picture the Rescue 2 guys hopping on the rig, heading into Manhattan. *This is the Super Bowl fire. This is the big one.* No one wants to miss it. Everybody in the firehouse, on duty and off duty, is on that truck right now, barreling toward the towers.

PHIL RUVOLO IS SITTING on his back porch when he hears his scanner get busy. Voices are coming fast and furious with news of a plane that hit the World Trade Center. Ruvolo has just finished three weeks of summer vacation and is slated to work tonight. The first thing that runs through his mind isn't tragedy; it's envy. *This is going to be one of those jobs that the guys will talk about for years to come in the kitchen. I'm missing it. Shit.*

"Oh, yeah, I was at the big one," they'll be saying next week, next year, next decade. "The World Trade Center, when the plane hit." Like most New Yorkers, Ruvolo figures it's a Cessna or a Piper Cub, something small that strayed off its flight path and struck the building. He doesn't get too excited about it.

Since Ruvolo came to Rescue 2, he and Bob have always managed to work the big ones: like last year's gas explosion down on State Street or the Atlantic Avenue Fire. But today, he thinks ruefully, their luck - hasn't held. Neither one of them is getting a piece of the action. So Ruvolo picks up the phone and calls Bob at home to complain. When Bob's answering machine picks up, Ruvolo leaves a message regretting that they're not in Manhattan, side by side, pushing up the stairs with a hose line.

Ruvolo hangs up and glances over at his television just in time to see a huge commercial plane hit the second tower. This stops him cold for a second. Ruvolo feels it in his gut: this has to be terrorism. Now he can think only of one thing: his guys in Manhattan.

JUST AS BOB gets on the Jackie Robinson Parkway, he hears over the scanner that a second plane has hit the South Tower. A chief on the scene tells the Manhattan dispatcher to recall all off-duty FDNY firefighters. *A recall?* In Bob's twenty-three years on the job, no job has ever required every man in the department.

Traffic grinds to a halt after Bob's car climbs to the crest of a hill near the Cypress Avenue exit. At first he thinks it's an accident. Then he sees that drivers have pulled over to the side of the road just to sit and stare. Bob is cursing those people, screaming, on the verge of some bad road rage, when he looks out the window himself. There they are, the Twin Towers, still crowning the city but now engulfed in smoke and flames. Bob can't even count how many floors are going. In his time, Bob has seen entire blocks burn down to the ground. But looking across the river this morning, he sees more fire than he has ever faced in his life.

Bob is the first off-duty man to Rescue 2, and he's there all alone for a few minutes, while he grabs his equipment and prepares to head in to the trade center. If he leaves now, he'll be a little behind the Rescue 2 rig, but he'll definitely get to do some firefighting. He calls over to SOC headquarters to tell them that he's going over there direct, by himself.

The fireman who answers the phone at SOC is adamant: "No, no. You can't go. Call the other guys and tell them to come in to work."

"No, man," says Bob. "I'm going in."

"No, you're not. You gotta make those phone calls."

Bob stomps up the stairs and enters the office. First, he checks the journal to see who is working. He already knows Billy Lake is driving, because he ran into him the night before at the change of tours. Billy was about to celebrate his twentieth anniversary in the department. He was treating the guys to a special meal. Looking at the book, Bob figures out who else is on duty: Pete Martin is the boss. John Napolitano, Kevin O'Rourke, Danny Libretti, and Eddie Rall are all working too. Lincoln Quappe's name isn't in the book as working this shift, but Bob saw his car in the parking lot. Lincoln must have hopped on.

Bob turns up the scanner volume and switches on the TV. Then he starts making phone calls. Most of the guys aren't at home; they're already on their way in. Soon they start pouring in: Paul Somin from nearby Middle Village, Queens, Ray Smith from Ronkonkoma, some men from Staten Island. The firehouse is filling up.

PHIL RUVOLO IS JUST EXITING from the BQE at Forty-eighth Street in Brooklyn when he sees the South Tower come down. It comes down in slow motion, the kind of thing he's only ever seen before in big-budget Hollywood movies. All of lower Manhattan seems to be engulfed in smoke, and he can't even tell if any of the tower is still standing. When he sees the tower go down, Ruvolo knows he just lost a lot of friends, not to mention a tremendous number of civilians. *It won't be my guys,* he thinks to himself. *Not my men.* He is certain of this.

He speeds through the borough now. The streets have a feeling of panic. Groups of people stand together outside on the corners that have a view. Others congregate in front of electronics stores, staring at the televisions, mesmerized by the television news reports. Ruvolo looks northward to the city whenever he can, but is only able to steal glimpses of the towers, and he still can't figure out what is left standing in Manhattan.

The street in front of the firehouse is filled with cars. They're all over the sidewalks, across the street, bursting out of the lot. The only time they ever have this many vehicles is at the Christmas party. Ruvolo walks into the kitchen and finds his men staring at the TV as they prep their gear.

Ruvolo doesn't have a lot of time, so he gets right to the point. "We're gonna go down there and do the best we can," he says. "We all lost friends today. But there will be survivors." Then he tells the guys to go downstairs and grab what they can: first-aid gear, hand tools, slings, and grip hoists. He wants them to be able to tend to victims and work the collapse zone, moving debris. But they have no radios and no masks; all of their most important equipment is on the rig, somewhere underneath the towers.

When he goes up to his office, looking at the journal and getting his own gear, Ruvolo does what everyone else in the firehouse is doing: he listens to the department radio, trying to understand what is happening. Normally, in this kind of situation, the chiefs would be all over the airwaves coordinating the response. But all Ruvolo and his men hear are line units in trouble, transmitting the kinds of messages no one ever wants to listen to. Messages like Brian Fahey's message last June, only worse, because there are so many more people involved. "I'm trapped," they hear. "I can't breathe." And they hear firemen begging for their lives.

Eighteen men are assembled at Rescue 2, almost every current member not already at the trade center, plus one retiree—Richie

Evers, of course. Richie has gone out on three-quarters salary, because of his bad lungs, but today he doesn't care about his lungs. He's going with the guys no matter what. The company he spent twenty years of his life building is in there. Billy Lake, whose son Richie is godfather to, is on duty at the site. Pete Martin, who was a fireman here with Richie for a decade, is down there. Lincoln, Richie's final protégé, is there. No way Richie is sitting this out.

A Chevy Suburban screeches to a stop in front of the firehouse, sent by SOC for the journey into Manhattan. But it can only take six passengers. Luckily, a city bus route goes right in front of the firehouse. No fireman has ever taken this bus, but now Paul Somin and Tommy Donnelly go out and wait for the bus. When it glides down Bergen, coasting to a stop near the corner, Paul and Tommy wave it down. They climb aboard and ask the bus driver if she'll take all of them to the Twin Towers. "No problem," she says. And she orders everyone else off of the bus.

As they step onto the bus, the Rescue 2 firemen take a quick look at the smoke on top of Manhattan, which is visible over the grammar school across the street. It's more smoke than they've ever seen, put in sharp relief by a summer sky free of any clouds—except the one over Manhattan.

Phil Ruvolo rides shotgun in the suburban sent from SOC, in front of the bus, using the lights and siren to clear a path for his vehicle and the enormous city bus that trails behind him. He strains to catch every word on the radio scanner, trying to make out what's going on. But the radio is now eerily silent after the dispatcher's calls for companies on scene at the trade center.

Somebody radios the dispatcher looking for the SOC command post. The dispatcher tries to raise someone from SOC: *"Manhattan to SOC command post."* Silence. *"Manhattan to SOC Command Post."* More silence. *"At this time, we are unable to contact the SOC Command Post."*

Ruvolo radios the Manhattan dispatcher. *"Rescue 2 second section to Manhattan."*

*"Go ahead, Rescue 2."*

*"We have eighteen men and we are heading over the Brooklyn Bridge. We'll report to the Park Row staging area."*

CHIEF DOWNEY ALWAYS GETS in to work early. His morning ritual is as consistent as clockwork. Arriving at his office at SOC headquarters, he heads out for a jog around Roosevelt Island while everybody else is still shooting the shit in the kitchen. When he comes back, he showers, then hits the kitchen. His appearance is the signal to the others that it's time to wrap up the conversations and get to work. As they scatter, Downey toasts a piece of raisin bread, spreads some jam on it, cuts it up, and then sits down to read the paper. That's the ritual.

This morning, however, his routine is interrupted by a code he hasn't heard too much lately: 10-60—a major emergency. Before the radio dispatcher even finishes announcing that a plane has struck the World Trade Center, Downey has one foot out the door. On the drive into Manhattan, coming across the Fifty-ninth Street Bridge, Downey and his chauffeur can see the smoke billowing up over the tall buildings of midtown. "This is gonna be the big one," Downey says to his - driver. "We're gonna be here for a while."

By 9:20 AM, Ray Downey is standing on the corner of Vesey and West Streets where the chiefs in charge have gathered to survey the scene. Since the second plane hit, at 9:03 AM, the scene has become even more chaotic. Downey checks in with the chiefs, who are trying to formulate a plan to deal with the chaos. Downey says he is worried about secondary devices in the towers, explosive devices that could hurt the firemen.

All five of his rescue companies are responding—by now, all five are either at work in the towers or on their way in. Six out of seven of his squad companies have also been assigned to the box. When Commissioner Von Essen comes over, Downey says, "You know these buildings *can* collapse."

Up above, in the North Tower, many of Downey's old Rescue 2 men are climbing the stairs. Terry Hatton, captain of Rescue 1, and

Dennis Mojica, his lieutenant, are leading nine Rescue 1 men up a stairwell. Patty Brown, another Rescue 2 eighties alum, is captaining Ladder 3. Timmy Stackpole, who Lincoln Quappe and Billy Esposito pulled out of the Atlantic Avenue collapse, isn't supposed to resume active duty until next week, but today he responded from headquarters. Mike Esposito, Downey's "illegitimate son," is in the towers as well, leading Squad 1. And of course, Downey's favorites, Rescue 2, are also climbing the stairs. All told, Downey has almost one hundred men directly under his command hitting the stairs or elevators, trying to get people out.

Captain Al Fuentes leads the Marine Division, the fireboats. He's already pulled all his boats into position near the World Financial Center, ready to evacuate the wounded or pump water to put out the fire. On his way to the trade center, coming around the southern tip of Manhattan by boat, Al saw a plane buzz overhead. *Holy shit! It's low.* He heard the screech just above him—the sound of the pilot gunning the engine. Then Al watched in horror as the plane hit the South Tower and exploded into flames. Now, as Al's men steer his boat into a berth at the World Financial Center, he says to them, "We're in a war here."

After docking his ship, Al walks east two blocks to the command post on West Street. He passes Chief of Department Pete Ganci, an old friend, talking with Deputy Commissioner Bill Feehan. Now he spots Downey, who gives Al a look that says it all. It says, *We're fucked, and stay with me.* It says, *Lots of people are going to die today.* And it says, *This is the worst thing we've ever seen in our lives.* Al has never seen Ray Downey look this way. He walks over to his friend because he knows Downey wants his help.

"Al," Downey says, "stay with me. Give me a hand."

"You got it, boss," says Al.

"Make sure that everybody coming in knows their orders," Downey tells Al.

The scene is unreal. Each time Al looks up, he sees something he can't bear to keep watching: people jumping from the highest floors of the North Tower, and then landing nearby. Each jumper sounds like a clap of thunder exploding on the ground.

As chief of department, Pete Ganci is running the show; this is his fire. Still, everyone keeps conferring with Downey to get his opinion on what he thinks is going to happen and what they should do with the men. They know they aren't going to be able to put this fire out. The best they can hope for is to evacuate as many people as possible. Looking up in the sky, they can see that a whole section of the building is engulfed in flames. One floor of a tower would be hard to extinguish, but this thing is impossible. Like trying to put out the Sun.

While Downey talks to the chiefs, Al meets each line unit as it arrives and makes sure that the men know their orders. "Al," Downey says over his shoulder. "Tell the guys to watch out for secondary devices."

Downey still thinks there might be bombs in there, or more planes headed their way. When he hears that a plane has hit the Pentagon, Downey confers with a guy from the mayor's Office of Emergency Management who confirms that another hijacked plane is still missing. Downey tells the guy to get on the fucking phone right now with the Air Force and get fighter pilots in the air to protect the city.

It's horrifying but Al can't stop looking up at the jumpers diving down to the ground. He can't stand just waiting here and watching these people kill themselves. He has to go up. He's ready to do anything that will get him away from the sounds of the jumpers.

"Ray," Al says, and he's almost begging. "Let me take a couple of companies up."

"No," says Downey. "You're staying with me."

Al walks over to talk to Kevin Dowdell, a Rescue 2 alum who's now a lieutenant in Rescue 4. He and his men are about to hit the stairs. "Kev, you know where you're goin'?"

"Yeah," says Dowdell. "I just need to talk to my men first."

The two towers are now spitting out smoke, ash, papers, and debris at a furious pace. Even for men who have faced down the biggest disasters in the city and in the country, the sight of these two enormous buildings burning out of control is fearsome. The most frustrating thing is to look up to the buildings and realize that those fires are not going to be put out. Still, in the face of this incomprehensible danger, confronting a situation that they cannot control,

Downey and the other leaders of the FDNY talk calmly to one another, with the same matter-of-fact attitude they would use at a big brownstone fire in Brooklyn. Fourteen years of infernos, disasters, and mayhem, fourteen years of racing across Brooklyn, commanding the company he loved, have made Downey unflappable. In his heart of hearts he must know that this one is different, that even the happiest ending is going to be utter tragedy. But today, like everyday, he isn't wearing his heart on his sleeve.

Al Fuentes and Ray Downey are still on West Street, across the street from the Towers, in a driveway that leads under the World Financial Center building behind them. In an instant Al realizes that something more horrible than the apocalypse he has already seen is about to happen. The sound is earth-shattering, and it precedes by only a few seconds the darkness of ash, dust, and debris that suddenly transform a sunny morning into night. He can't see anything at all. He sprints away from the collapse and runs down the driveway, into the building opposite the World Trade Center. He can't see his hand in front of his face. He doesn't know if he is about to die in this collapse too. He just huddles down in a corner and waits, prays, for some way out. A few minutes later the air starts to clear a little and Al gets up and walks tentatively out of the garage. Now he sees Ray Downey.

"We just lost a lot of good men," Downey says to Al. He is crestfallen, destroyed, worse than Al has ever seen him in his life.

Three months ago Al and Downey had stood on the site of the Father's Day Fire and Al had watched as his friend tried to take it all in. Ten years ago, Al had sat next to Ray at Potter's Pub and Downey had talked about the death of a probie, a boy he had tried to save but couldn't. Downey had cried. How could he bear this? How could anyone? But there's no time to grieve, no time to feel anything but shock and adrenaline. It isn't over yet. They are still under attack. More planes might be coming.

"Ray," says Al. "Let's start searching."

Now, as Bobby and Ruvolo race toward Manhattan, Downey and Al are walking across this landscape of ruin. The other people they see are running away, but they walk toward the disaster. There is nothing to do now but act. They'll be months, years, the rest of a lifetime to think. A chief tells them there are firemen trapped in the Marriott Hotel, next to the South Tower that just collapsed. The North Tower is still burning.

"Where the fuck are those fighter planes?" Downey asks Al, as they make their way to the Marriott.

They can see a group of chiefs and firefighters standing inside what's left of the hotel. The people in the group look injured and dazed. Downey tells Al to stand on West Street and signal when it's safe for people to come out. Safe? It's not ever going to be *safe*. But Downey makes a run for the hotel. Al stands there looking up at the debris falling from above. Inside the rubble of the Marriott, Ray Downey finds his old friend Brian O'Flaherty. O'Flaherty was captain of Rescue 1 in the era when Downey captained Rescue 2; now in this godforsaken place they meet again. O'Flaherty has a serious shoulder injury, so Downey helps him to climb over a wall of rubble.

Downey signals to Al that he wants to send O'Flaherty out. Al gives an all-clear wave, and O'Flaherty runs out, struggling with his painful shoulder. When two other firefighters climb out of the rubble— both of them from Squad 41 in the Bronx—Downey tells them to get the hell out of there.

"Go north toward Chief Ganci," he says. Al waves the two guys out.

When that first building came down, Al was sure that was the worst sound he would ever hear in his life. But this time, when the north tower collapses, it's much much worse. Al is closer to the building and this time he is simply buried under debris, unable to get out, to move, or to breathe. When the tower falls, Al is right on the periphery of the collapse. A few feet closer to the building and everyone seems to be dead. A few feet away and they are running for their lives. That second collapse is Al's last conscious memory of the day.

But somehow Al is functioning. One of his lungs is collapsing, he has horrible head injuries, and he's coming in and out of consciousness. Buried and completely pinned, nearly dead, Al has the presence of mind to dig out his radio and get on the Manhattan frequency.

"*Captain Fuentes to Manhattan,*" Al says. "*We have guys trapped in the collapse.*"

"*Manhattan to Captain Fuentes,*" the dispatcher responds. "*Cap, can you tell us where you are?"* Al is disoriented and incoherent. He doesn't respond.

THE SUBURBAN AND THE BUS carry the second wave of Rescue 2 men across the Brooklyn Bridge into downtown Manhattan. On the bus ride over, every man looks out the window. As they cross into Manhattan on the bridge, the second tower comes down. The men know that something has collapsed, but it is so smoky, dusty, and dark that they cannot see what remains standing.

All Ruvolo can see is an enormous dust cloud. He has no idea of the extent of the collapse. Did the tower fall straight down or lean over? How many other buildings are involved? Were the firemen in the tower told to evacuate? Did they have time to get out?

*A large number of members are gonna be dead,* Ruvolo thinks. *A large number of civilians are gonna be dead. But it's not gonna be all my guys.* He is certain of this. *We're gonna find a void. They'll be in it. They'll be beat up, but they'll be all right.*

On the radio, Ruvolo hears the dispatcher calling for command posts that used to be down here. Again, in response, there is only silence. The dispatcher tries to raise Field Com, the communications unit sent when this first happened, but there's no response. The sky goes so dark that the bus driver has to stop right by city hall; she can't see well enough to get them any closer. So the guys disembark. The men stay close together.

Ruvolo sees a chief just south of city hall and the captain reports in, giving the names of all his men to the chief. The chief has set up a command post right in the middle of the street, just in front of J & R Music World. The chief orders Ruvolo to hold his men there, until he can record everyone's name.

The guys are chomping at the bit. *We're like starving people at the buffet line,* thinks Bob. *Just let us go.* Ruvolo feels the same way. They are still blocks away from the towers, but visibility is only fifteen or twenty feet, so they have no idea of the extent of the collapse yet. They want to get down there and start pulling guys out.

Finally they get the okay to go. Ruvolo splits the men into three teams, each with one officer in charge. Ruvolo takes his group toward West Street, Lieutenant Pete Lund works his way toward Broadway, and Lieutenant George Hosle veers off toward the 10 Engine and 10 Truck firehouse on Liberty, just south of the trade center.

Ruvolo's team sets out down Broadway, passing by the trade center site, then turning right onto one of the little alleyways just south of the towers. It is nighttime in the middle of the day. As they come within a two-block radius of the collapse, the dust and smoke make it almost impossible to breathe. Bob spots a couple of guys smashing in the window of a jewelry store. Under normal circumstances he would chase them down, but now he just yells at them, and they stop, at least until he passes.

As they approach West Street, about to turn north toward the towers, the guys sink into water up to their knees, stumbling through floating debris. As he turns the corner onto West Street, Bob can't believe what he is seeing. *This is a movie set. This can't be for real.* He sees fire trucks on fire, engines burning, police cars tossed around like potato chips, ambulances crushed, cars flattened like pancakes. There is nothing that he's ever done in his life that could prepare him for what he sees when he turns the corner and gets a good look at what is—and what isn't—there. The buildings don't seem to be standing anymore, at least not that Bob can see. There are enormous piles of twisted steel everywhere. The air chokes and blinds.

Ruvolo finds it difficult to see more than a few feet ahead. He doesn't know where the buildings start and end. He doesn't know where he is anymore. He knows he's on rubble. How much rubble, he can't be sure. But immediately all his men fan out in different directions, trying to find the good void, where the guys are trapped. Even in this horrific landscape, their instinct never wavers: *Get the men out.*

Ruvolo is familiar with the WTC. In 1982 he worked at 10 and 10, just across the street from the towers. In 1993, when the first WTC attack occurred, he responded with Rescue 2. Though he has been right here hundreds of times before, he now has trouble orienting himself.

Ruvolo has been to dozens of collapses, but this is a thousand times bigger than any of them. Questions race through his mind. *What units are in what building? Are people trapped down in the parking garages or the PATH tubes? Is another plane gonna hit us?* Then he realizes: he can't worry about all these complications. Right now he just needs to find a foothold here, somewhere to start the operation.

He can make out pieces of what he thinks was the Marriott hotel in the rubble in front of him. That's where they'll start the search. This is a little more familiar than what he just came through. It's the kind of collapse the men are used to. Debris, structural parts of the building, wood, steel, papers, remnants of carpet are all smashed together. There are some places to search, voids to climb into and explore. They find a few dead bodies in the rubble. That is just what they expected; it doesn't phase them. For now, they don't move the dead. They just keep looking for the live ones. That's why they're here. There will be time for the dead later.

Ruvolo realistically expects to find hundreds, maybe thousands of bodies; some dead, some alive. His men are hauling all the first-aid equipment they can carry, thinking that they'll be patching people up all day. What is most upsetting is that, even when they venture further into the collapse, they don't see many bodies. They keep thinking they are going to find a huge pile of people, but they don't—at least not here, in the ruins of the Marriott. Even though there aren't many bodies, the feeling of death is all around them.

Every off-duty Rescue 2 guy is somewhere on the pile, doing searches. Retired men from throughout the city have headed in too. Pete Bondy has been out for ten years, but that hasn't stopped him from speeding in from Long Beach and diving into the rubble. He and a chief he knows just pushed past a deputy chief and began searching—like all the rest of the guys. Bondy finds some dead firemen and hopes he's close to finding a fireman with a pulse. Another group of searchers is trying to figure out how to get one guy, who is dead, out of a void; Bondy volunteers to climb under and free him up.

Some of the retired guys are just here to lend a hand, but others are searching for their sons. At one point Bondy runs into Lee Ielpi, another retired Rescue 2 guy whose son Jonathan is missing. Pete takes one look at Lee and decides to drive him home immediately. John Vigiano, a longtime Rescue 2 lieutenant, is looking for his son from Ladder 132 and his other son, who works as an Emergency Services Unit cop. It also works the other way: Ray Downey's two sons are scouring the pile for their dad.

Over at Ruvolo's corner of the pile, one phase is ending. The men have hit all the voids they can get to in the collapsed hotel, and now Ruvolo leads them north, toward the towers. It's hard going on West Street because the street is packed with debris. To move three hundred feet takes forty-five minutes of climbing and crawling.

Along the way Bob spots some firemen digging a guy out and goes to help. They are about to lift him out when Bob, fresh out of his EMT refresher course, says, "Woo, bro, no. Let's get a backboard. He survived this. Let's not kill him." The guy is smashed up, but it's mostly cosmetic. He'll live.

Now Bob looks around for the Rescue 2 guys but they're gone. After the adrenaline of the rescue wears off, he has trouble breathing. He's perspiring. *I'm having a heart attack,* he thinks. He spots the remains of the Rescue 2 rig, across West Street, and stumbles over there. The rig, a battered shell of its former self, is empty now. Every piece of gear has been taken off the shelves and must be somewhere there in the rubble. Somehow the sight of this broken-down rig, the rig they dedicated to Louis just a few years ago, breaks Bob's heart. Every compartment is open and bare. Richie is sitting at the rig, clearly in horrible pain. After neck surgery that he hadn't even fully recovered from, Richie has popped his neck again here, searching in the rubble.

"Richie," Bob says, "I think I'm having a heart attack,"

Bob sits on the bumper of the rig for a few minutes and feels every beat of his heart pounding in his chest cavity. It still hurts, and his breathing is tight. Then Bob thinks, *What's wrong with me? I'm pussying out. If I drop dead of a heart attack, I drop dead. I don't care. I have to go back out there.*

Leaving Richie at the rig, Bob goes back across West Street, where he meets up with Paul Somin. They circle the entire site, looking for a good way into the pile. There isn't one. Eventually they just tackle it head on, entering through a valley formed between the North Tower and West Street.

Bob is somewhere between the two towers, sifting through the ash, looking for anything he can recognize. He looks down at the stuff underneath him. He looks at the empty hulks of buildings surrounding him. If this collapse can mangle a two-ton steel rig parked across the street, what must it do to a human body? And he thinks, *Holy shit, everybody's dead and we gotta find 'em. Oh man, John's dead. Oh man, Lincoln's dead. They're all dead.*

Then he says it out loud. "They're all dead."

"What're you, nuts?" somebody yells at him.

"Bro," Bob says sadly, "they're dead."

Next he thinks of the families sitting at home watching the television coverage of this disaster, calling the firehouse, calling headquarters, desperately trying to figure out what happened to their men. *Holy shit. What do we tell the families? Working down here in the heat and flames, coughing up a lung, dangling from an I beam a hundred feet off the ground, this is the easy part. This is what we know how to do. But what are we gonna say to the wives?*

At this moment Ruvolo and what is left of his team, who have been walking north up West Street, come to a stop in an area that must have been right between the two towers. But this collapse zone is something entirely different from the others they've passed through. The heat below-grade is tremendous. It's like being in a fire, where you can see just a few feet ahead.

There are no visible signs of what used to be the towers except the enormous steel I beams. The I beams are red hot, so hot that they'll burn you if you touch them. Everything around them is just pulverized ash. And it's all painfully hot. Visibility is still minimal. There is molten metal streaming down the channels in this collapse. The carbon monoxide levels are through the roof. But nobody has a tank to breathe with, those too are somewhere in there with the first wave of Rescue 2 men. Now the second wave of men dart into the voids quickly, so they don't suffocate. But each void they find is empty.

Normally, collapse work means removing debris and working in a little further. Then you shore up behind you, so a secondary collapse doesn't get you. In this case, there is hardly any debris that Ruvolo and his men can remove without a crane. The steelwork that remains is five inches thick. They can't cut that or drag it out. They have to work around it.

Since the guys don't have masks, their eyes cake up with dirt, ash, and soot almost immediately. Then they start to tear, swell, and turn bright red. After a few minutes nobody can see anymore and the only thing to do is to try to find some water and wash out their eyes. The men fan out across this field of ash, avoiding the pockets of fire and being careful not to touch the molten I beams. Tough old Larry Senzel makes it all the way in to the east, toward the towers. This time, however, they don't even find any dead bodies, never mind live surface victims. After an hour at this location, between the towers, just off West Street, there are no more voids left to search. Ruvolo's team trudges back out to West Street and keeps working it's way north.

Ruvolo thinks they are now at the southwest corner of the North Tower, but he still doesn't know the extent of the collapses. It's so cloudy that for all he knows the bottom twenty-five stories of each building might still be standing; there's no way to tell. The pile reaches up many stories. There are pieces of the outer shell of the building, enormous steel hunks; there is more of the same compressed ash and concrete particles. Searching here is even more difficult because they have to climb a mountain to do their work. Here they meet up with other firemen at this location, who share the latest news from the gossip they've heard on the pile.

Every fireman who comes by asks and is asked, "What have you heard?" Fairly quickly, Ruvolo hears about the chiefs and top brass who are missing: Pete Ganci, Bill Feehan, Father Mychal Judge. The upper ranks of the department—the people who would normally be in charge here—are down in the rubble. This would be Downey's operation, no doubt, but Ruvolo hasn't seen Downey or heard him on the radio. Can Downey himself be missing?

Word starts to filter in about the rescues and squads. All five rescues had responded initially, as had six of the seven squad companies and the haz-mat unit. All these men were under Downey's command. Ray Downey's men should now be the ones leading this search-and-rescue operation. But not only are a lot of his men probably dead—the special equipment that his rescues and squads carry has been lost in the collapses. A few of the rigs have survived because they were parked far away from the scene, but most, like the Rescue 2 rig, have been destroyed beyond repair.

Ruvolo hears a lot of general information, and some reliable-sounding reports about a few deaths. But he hears nothing definite about the subject that consumes him: the fate of his own men. Where are they? Are they alive somewhere in here? Evacuated to a hospital in New Jersey? At work on the other side of the pile? He just doesn't know. And communication has broken down to the point that he may not know for hours—even days.

Every one of the Rescue 2 men has done the calculations. The guys left quarters at around 8:50 AM. They probably arrived here at 9:05 or 9:10, just after the South Tower got hit. But since they were originally assigned when the North Tower was hit, odds are they went up into the North Tower—the last to fall, the one that guys might have evacuated from before it collapsed. But other than these hypotheses, Ruvolo knows nothing. He can't find anyone who can tell him what was happening here before the buildings came down. And every time he meets another fireman who only arrived after the collapses, it starts to seem like almost everyone in the first wave of firemen must be trapped in there somewhere.

They climb up the ruins of the North Tower any way they can manage, using every ladder and sling at their disposal. Ruvolo and his men are desperate to find their guys. They scale a hundred-foot-high mountain, always looking for shoes, arms, hands sticking out. The radios are frantic with Mayday messages, but few of the trapped men can say where they are. Jimmy Kiesling manages to locate a civilian who is trapped. Kiesling, the best rope rescue guy in the city, slices up a piece of webbing into smaller pieces, then rigs it into a sling to lower his victim down to safety.

The men of Rescue 2 are searching the ruins of what Ruvolo thinks was the North Tower. They climb girders, box beams, I beams, anything that will get them up higher. They pass ladders to the men in front, who place them across the wide gaps in the material they are climbing over. All of this is risky; Ruvolo knows how easy it would be for a rescuer to plummet to his death. But right now he's also willing to give them the freedom to go into dangerous places; he believes they still have a chance to pull out some live victims if they work fast.

Truth is, the guys are taking crazy risks up here. Safety is out the window. No one is footing the ladder. Ropes and ladders make a chain up the side of the collapse. The two sides of the frame above them are creaking. *It's coming down any second*, Bob thinks. Chiefs are calling them off of the pile. But nobody wants to punk out. Time doesn't mean anything.

The goal is to get into the stairways; that's where the people should be. Another chief shows up screaming at Rescue 2 that they have to get out. "Building Seven is going to collapse on us. Forty-seven stories. It's coming down!" And sure enough, the men can see it burning out of control, above them. No one would leave willingly, but they are also a little glad to be ordered off the pile.

Some other firemen manage to rescue the guys from 6 Truck who were holed up in a void with two civilians. That's the best news anyone has heard all day. Bob and the guys around him tell the men from 6 Truck to go ahead; they'll help one of the civilians get down. They help a big guy with an injured leg. He limps down the slope, hanging onto a fireman. When he gets to the bottom, he just sits down and says, "I gotta take a rest."

"A rest?" Bob asks. He shakes his head and helps the guy to his feet. "Not yet, bro."

Behind them is number 7 World Trade Center, now fully involved and about to come down on their heads. Before this morning, such a collapse would have been the largest in New York history. But, with two 110-story buildings down in front of them, the smaller building seems like nothing. Once they get out of the danger zone, everyone watches the building burn. Ruvolo and Pete Lund are sitting on the bumper of the Rescue 2 rig when someone comes along screaming, "It's gonna go. It's coming down."

"Are we safe here?" Ruvolo asks.

"Yeah," someone says.

Ruvolo doesn't even look up when it falls down to the ground. He can't be bothered. It doesn't even matter.

While they wait for the okay to go back onto the pile, the men have a few minutes to think. It's the first real breather most of them have had all day. Bob is relieved that he no longer feels like he's going to drop dead. He feels like shit, but the heart attack symptoms have passed. While he stands there and waits, taking a rest, he keeps hearing about others missing or confirmed dead. Bob, like all the other firemen, has a wide circle of friends in the department. There are the guys he works with now, in the rescues and squads. There are the guys he worked with in his older ladder company, the guys he was a probie with, the guy he was a volunteer with on Long Island, the guys he worked with at the Boardy Barn. One way or another he probably knew a thousand city firemen. He still has no idea which ones are alive and which ones are dead.

Then he starts thinking about this morning. If he had come here with the EMS chief, he'd be out there under the rubble, like the rest of the Rescue 2 guys. If he had come here immediately after he picked up his equipment at Rescue 2, he'd be out there under the rubble. If he had switched tours, juggled mutuals, changed groups, he would be out there under the rubble. *You were at the mercy of the group chart today,* he thinks. *Anything can happen on any given day. Don't want to think about it.*

Ruvolo is starting to cramp up everywhere. He recognizes that as a sign of dehydration, but can't find anything to drink. Finally he passes out in an ambulance. By nightfall, many men will have to go to an ambulance for an IV and an eyewash. Many guys will go twice. They know the feeling when they hit the wall. It's happened before in fires. Everything hurts, and you can't take another step. The paramedics in the ambulance don't want to let Ruvolo go back after they take his blood pressure, but he knows he has no choice. The only thing they can do is pump him up with saline. There's no water to drink, so Ruvolo knows he's gonna cramp up again and have to come back into the ambulance. When he does find a bottle of water, it's like a gift from God. He and Pete Romeo sit down and split it.

After a fruitless afternoon and evening of searching, Ruvolo prepares to reenter the outside world. At about midnight that night, Ruvolo and two groups of men take a boat back to Brooklyn. Lieutenant Lund remains overnight with his men. The bridges and tunnels are closed, so the boat is their only option. He dreads the calls he will have to make to the wives who are sitting at home worried sick over their husbands. Ruvolo has no news. No bodies have been found. They might be alive, or they might be dead. He just doesn't know. The confirmed deaths are few: a fireman hit by a falling body early on, before the towers collapsed; Father Mychal Judge; and a number of the top brass of the department, including Pete Ganci. Ray Downey was supposed to be with the men whose bodies have been found, but they haven't found his body. Ruvolo knows that Rosalie is at home, praying he's alive.

Normally, a whole team of men would go out and break the news to a waiting wife. Counseling units, priests, and chiefs would be overseeing the whole thing. But now it's just Ruvolo, alone on the telephone line. He could go out and see each wife right now, but he'd rather take the time to try to bring their husbands back alive. An injured Rescue 2 fireman, Cliff Pase, has been manning the phones all day. Now Ruvolo has to talk to the wives. It is by far the worst set of calls he has ever made in his life. Every wife is just waiting for the firefighters to pull her husband out alive. Every wife tells him that is going to happen. Ruvolo, who has seen the reality of that pile, has less hope than they do.

As usual, Ann Napolitano talked to John on the phone on the night of September 10. He was excited about a special dinner they were having at the firehouse, for Billy Lake's twentieth anniversary with the department. John would never say that he wanted to get back to the guys, but Anne could hear in his voice that he was eager to get back to the dinner. She could hear laughter in the background, and she was happy that her husband, after trying so hard for so long, was finally one of the guys.

When she sees the news the next day on television at work, she heads home as soon as she can. She's afraid to pick up her youngest daughter from the day care center, because she feels too shaky to drive. But the staff insists that she take her daughter; they're closing for the day. At home she gets on the phone and starts calling the firehouse. No news. The men are missing. That's it.

All day, John's father has been calling the house and getting a busy signal. John must be home, his father figures, because Anne is at work. When he gets through, his worst fear is realized: John was working this morning. After that, John's father just can't stand to wait at home. He makes his way into the city. When he reaches the site, he dodges a set of police barriers, using his old police credentials. No one has any word of John. Near a triage center he scrawls a message on the wall, in the ashes: *John Napolitano—Your father is looking for you.*

Later, some of the Rescue 2 guys see the message and look around for John's father. But John's father doesn't really want to be found. He knows the Rescue 2 guys would only sit him down and hold him back. He makes his way around the site, looking for his boy, hoping.

MARYANN O'ROURKE TALKED TO KEVIN the night before the planes hit. Normally she would have called him this morning, but she has to be at her nursing job for two surgeries on Tuesday morning and leaves home too early to call Kev. She is at work at the doctor's office when the first plane hits, but strangely enough, she doesn't think about it much. *Kevin's going. Of course he'll be there. But he'll be fine,* she tells herself.

Then her daughter Corinne calls from college upstate. By the time Maryann gets to the phone, Corinne is crying hysterically. Corinne has always been the strong kid, the one who doesn't show her emotions. Maryann starts to worry a little bit while she tries to console her daughter. But she still believes it herself when she says, "Daddy will be fine. He'll call us later." After Maryann hangs up with Corinne, her younger daughter Jamie calls from her high school, where the students are watching the news in all the classrooms. She can't bear to watch, knowing her father is in there. So Maryann calls her sister and arranges for her to bring Jamie home.

As soon as she can, Maryann heads home. She tells herself that Kev will call, but that it might be a while before he can get to a phone. The phone lines in the city are a mess and most cell phones aren't working. It's only on her way home that she hears that both towers are down and that firemen are missing. She calls Rescue 2 and no one answers. She calls the department's 800-Help number, but the guy there just says: "There are a lot of men missing and we're trying to figure out who right now. We'll call you as soon as we know."

About midnight Phil Ruvolo finally phones. "Kevin is missing," he says. And that changes everything for Maryann. But he can't tell her anything more.

LATER THAT FIRST NIGHT, after he returns to the firehouse exhausted and spent, Bob gets into his car and drives home to Long Island. He and Ray Smith, another Rescue 2 guy, want to visit all the wives who live near them, including Ann Napolitano, Jane Quappe, and Eddie Rall's wife, Darlene. It's already one or two in the morning, so their plan is to see if the lights are on before they knock on the doors. They go to John's house first, not far from where Bob lives. Bob feels shaky just walking up the driveway, worse than he did even when he nearly passed out at the trade center. Of course the lights are on; the lights are on all over. No one is sleeping. Bob has no idea what to say. He can't tell John's family what he feels: that John is gone, dead. He has to give hope, even though he doesn't feel it. So he does his best to comfort Ann and all the rest of the wives. But he will not say what they want to hear, if he thinks it is untrue. He will not say, "We're gonna bring him home alive." He just can't. That will only make it worse.

Bob can hardly look at Ann. Seeing it all in her face is the worst. *They need us to be steady,* he tells himself. *They need us to keep some kind of faith.*

"Do you pay your mortgage or does he?" Bob asks. "Do you need anything from the grocery store? Diapers, milk?" As soon as he can, he moves the conversation from the emotional to the practical. That's what he's good at. He'd rather pay bills or buy groceries than talk about feelings. Taking care of practical matters makes him think that his world is still under control.

Bob and the other firemen usually refer to the wives as "the girls." To the firemen it's not a condescending term; it's just what they call a group of women. Bob appreciates how loving and caring all the girls are, and they in turn appreciate him coming. They understand that he and the others are risking their lives to find their husbands. But Bob just doesn't know what to do with their kindness. Having to be emotional in front of the girls, or in front of anyone, is agony for the firemen. They don't do that with one another, not ever.

When Bob finally makes it back to his own house, Linda and his daughters just stare at him for a while, unable to believe that he's still alive. Then he does what every other living New York City fireman is doing that night: he tries to explain why, the next morning, he must go back to work at the pile and put himself in danger again. There is no real answer to the question "Why do you have to do this?" There never is. Not when Louis died, not tonight, and not in the future.

The only thing Bob can say is, "It's what I do." For the men, that is enough, but for the wives that is like no answer at all.

AFTER HE FINISHES his phone calls to the wives, Phil Ruvolo hits the bed in the corner of his office and tries to sleep. For a few minutes, before he nods off to sleep, he thinks about what happened today.

Losing even one man haunts an officer for life. Every day without fail you will think about the fire where he died, the anatomy of his death, and what could have been done to prevent it. If you're the captain, they're your boys, no matter how old they are. And even if you weren't working when somebody got killed, you'll still wonder whether you could have taught him something, sometime, that would have saved his life. Now Captain Ruvolo is facing an extreme that so far exceeds his worst expectations, it seems to strain the limits of his mind.

You can't play favorites among the dead, but there are pains of recollection that are starting to hurt so much that the captain knows they will never go away. He's thinking of everyone and no one at first, trying to get his mind around the idea of a firehouse without seven of his guys. Then the guys start to come to mind, and he thinks of John Napolitano, the good son, the youngest member of Rescue 2 at thirty-two. Captain Ruvolo saw himself at Rescue 2 for a good ten years more, but after that he wanted John to take over as captain. Just two months ago John walked into the kitchen, so proud of that new tattoo of Rescue 2's logo, an angry bulldog, on his bicep.

Then there are more fragments of a life together that is now lost. The firehouse is an especially warm and welcoming place on a winter morning. The captain, like all of his men, always arrives at least two hours early. He often brings a box of muffins or pastries from the bakery near his home and tells the guys, "Try that bran muffin with some cream cheese," or "That *sflogliatelle* is amazing. Come on, you gotta have some." The two hours before his shift are spent in the kitchen talking with the night tour, hearing about the fires he missed, the meals they had without him. And each member fits perfectly into this atmosphere: John Napolitano, Kevin O'Rourke, Lincoln Quappe, Billy Lake, Danny Libretti, Eddie Rall, and Lieutenant Pete Martin. They are missing seven—*seven*—very good men. And now Ruvolo knows that when he walks into the Rescue 2 kitchen, he'll never have that feeling again.

# NINE:
## RECOVERY

WHEN THE RESCUE 2 MEN RETURN to the site on the morning of September 12, they begin with the enormous pile that was once the North Tower. Paul Somin and Billy Esposito are soon at the head of the group, leading the effort to scale the tremendous heap of steel, ash, and concrete. Some chiefs down below holler up at the guys to get off the pile, but the men ignore them. This could be where the guys are trapped. Rescue 2 is determined to go through each and every void to try and find anyone. They're hoping for survivors. Mostly they have pulled dead bodies off of this pile. Yesterday, however, Jimmy Kiesling helped get a survivor out from here. And this is the place where the men of 6 Truck walked out. You never know.

Moving carefully through the debris, Paul and Billy make it up to a section that was once the twelfth floor of the North Tower. They can tell it was the twelfth floor by the markings in a piece of the elevator shaft. All the way up the pile they yell, "Hello, hello, anybody here?" hoping for a response, trying to rouse someone. But they hear nothing, just the eerie quiet they can't get away from. Near the top of a 140-foot pile Paul bellows, "Hello," once more, at the top of his lungs. It comes out raspy and hoarse—his throat is clogged with ash and dust.

Suddenly he is yelling at the other firemen. "Everyone shut up!" he screams at the guys. "I think I just heard someone." The other guys are skeptical; they have to be. By now, twenty-seven hours after the towers collapsed, they've heard hundreds of false alarms. But they also have hope that this time it won't be the wind, or the building, or another rescuer making the noise. So they quiet down and let Paul call out again.

*Am I hallucinating?* he asks himself. *Did I really just hear that, or is it just exhaustion talking?*

"Hello," he yells out. "Anybody in here?"

This time the response is unmistakable. "Help me," he hears. A woman's voice. He's got someone. He just has to find her.

Edging over toward a void in the pile, he finds a partly intact stairwell. Hoping like hell that it will hold him, he ducks in, going under the surface of the pile, and sprints down the stairs to the eleventh floor. "Hello!" he yells out. "Hello, where are you?"

"I'm here," she says, but her voice is fainter now, so he runs up two flights of stairs and screams again.

"Hey! I'm right here," he hears, much louder this time.

"We got a survivor! We got a survivor!" he yells out to the firemen nearby.

Larry Gray, their former lieutenant, who has come out of retirement to work alongside his men, gets on the radio. It is September 12, 2001, at 12:30 PM. *"Rescue 2 to battalion. Mayday! Mayday! We've located a survivor in Stairwell B of the North Tower. We're on about the thirteenth floor."*

Across the enormous site, every person near a radio stops what they're doing for a minute and just listens. Firemen nearby rush to lend a hand to the rescuers. Word goes out around the pile: "They got someone. Rescue 2 got someone, over in the North Tower."

Meanwhile, Paul is still trying to pinpoint the woman's location. He can hear her fine, so he knows she's close, but he still can't see her in all this rubble. "What's your name?" he asks.

"Genelle."

"Genelle, we're not gonna leave you. We're gonna get you out. Can you stick your hand out so we can find you?" Paul asks. She has to be near. She sounds close by. But he still doesn't see anything. "Espo, you see her?" he asks Billy Esposito. "I don't see anything."

Then Espo sees her. A hand is sticking out of a pile of Sheetrock. At first he missed it because the fingers are dusty and gray, like everything here. Espo grabs her hand. They've got her.

"We found her," Paul yells. "Up here. We found her."

No less than twelve stories of building and debris separate them from the ground. The air is smoky and pungent with jet fuel. Nowhere they stand feels solid; the pile constantly shifts under their feet. They have no safety lines to hold them as they work. But right now Paul and Espo aren't thinking about any of that. *We found someone,* Paul thinks. *She's alive. There are people alive here. Our guys might be alive in here.*

"*Rescue 2 to Battalion,*" Larry Gray says again. "*That's a confirmed survivor up here, in the North Tower. We've located her and we are going to dig her out.*"

All the firemen nearby converge on the scene. Espo starts digging. Paul takes the woman's hand. Larry Senzel and Espo work from the top, trying to excavate a victim who can move nothing but a hand. She is pinned beneath a dense pile of Sheetrock, concrete, and metal.

"Thank God you found me. I've been up here for a day," she says.

"Genelle, don't worry," Paul says, speaking calmly now. "We're getting you out."

"What's your name?" she asks.

"Paul," he replies.

Everyone works to free her—*steady, careful.* The pile is slippery and unstable.

People are still shouting, "We have a survivor! We have a survivor!"

"Where were you when the building collapsed?" Paul asks the woman.

"I was with a group of people on the thirteenth floor."

They spend half an hour digging her out. It is exhilarating—the best work they have done since they came to the pile. Working with gloved hands, they rip apart Sheetrock, pry metal, rapidly unpack dense pockets of debris. They try anything to maneuver her out. Finally, somebody arrives with a cutting torch and starts to work on the metal around her. First her arm is uncovered and she can still move it. They work backward from the arm toward the rest of her body. Finally, there are so many men at work that she begins to be freed in multiple places at once. She can wiggle a leg, then both legs. Her arm, then her upper body. And then, after what seems like hours, she is free. Gently, the men lift her out of the place where she expected to die.

The woman's eyes are swollen shut, and her face is terribly swollen too. Seeing her, Paul thinks she is obese. Her hands are enormous. But in the days to come when he sees her photo in the newspaper, he'll realize that she is actually petite. It was just that that the injuries she sustained made her body swell a tremendous amount.

They bundle her up in a Stokes Basket, tying the seat belts on top to strap her in, and pass her down to a team of waiting firemen. But they don't stop searching after they send her away. They don't take a breather or go down for a rest. There is no time to stop and pat one another on the back or go down to talk with Genelle and make sure she's okay. In fact, the firemen never see her again. In the days to come, Paul will read a couple of interviews in which Genelle mentions a firefighter named Paul who found her and held her hand, whom she wants to contact. But he won't get in touch with her. That's not his way of doing things. He doesn't want the glory and, he thinks, he doesn't deserve it. They all deserve it, everyone in the FDNY, so it's better like this.

Now, as the guys get ready to search, dig, and closely examine everything in this pile, they think, *This changes everything. Now we know. People are still alive in here.* They are ready to take even more risks, to find the civilians and the firefighters waiting for them under this pile.

"Paul, your big mouth finally came in handy," says Larry Gray. And the guys laugh for a few seconds before they start searching again.

The next person they find after Genelle is dead. They can see his captain's bars and read his name tag: Captain Hynes from 13 Truck. After marking his body they move on to look for survivors. But they just find more dead bodies, mostly firemen, on the outer edges of this collapse, the sloping walls of the pile. More dead. They continue to dig furiously, marking the dead as they go, hoping to find someone alive.

Genelle lasted more than a day in this pile. Out of all the thousands of other people who got caught in the collapse, somebody else has to have made it through these twenty-seven hours. Or so the men reason. But, after hours of hard labor, combing through every possible pocket or void they can find, yelling for victims until their voices are shot, the men find only more fatalities. Or pieces of fatalities. Finally, they work their way down the unstable slope of the pile to take a rest. This day that started with exhilaration is now ending with utter disappointment. It is still Wednesday. But in the smoke and gray air it seems like no day in no place they have ever known.

BOB AND THE OTHERS take a rest over by the tool cache, where the FDNY has stockpiled equipment; they lie down in the dirt and rubble, trying to stave off total exhaustion. At some point a chief calls for them to come and look at a list. On the document are the names of everybody whom the Fire Department thinks is missing. The number of men who have not been located is staggering. The names go on for page after page. Bob recognizes many of the lost right away. In some instances, he knows the name but can't picture the guy.

After just one page, he has read the names of so many dead people he knows that it is hard to keep track of them. So many days and nights and years flash before him. The scope of this is unreal, incomprehensible. There are simply too many names of dead people he knows to remember them all. Or to begin to mourn or feel at all.

This is just a preliminary count, which the chief is showing in order to see if anyone has seen any of these men alive. Spying one of the names, Bob is actually able to say, "I just saw that guy. He's right over there." When the chief is able to cross one name off the long list, a little bit of relief comes into his worn-out, dirty face. Bob feels that same thing. For a moment. But, as his eyes continue down the list of names, he is not able to find anyone else he has seen alive. He thinks to himself that, even if a lot more names come off, that is still going to be one very long list of dead brothers, more line-of-duty deaths than he ever imagined he would see in his entire life, never mind in one day.

The next couple of days are a blur. The men work eighteen or twenty hours a day, sifting through the rubble, crawling into voids. Sometimes they catch a few hours of sleep on the street. They don't go home, even to sleep, and the civilian world seems as distant as another planet to them. These days, they don't hear any more sur- vivors crying out from the pile. There is only the smoke and the smell. They find plenty of dead people. And they begin to wonder if Genelle Guzman will be the last person to make it out alive.

ON SATURDAY, Ruvolo speaks to Lynn Tierney, a deputy fire com- missioner, and to a chief's aide. Both were at the site from the begin- ning on the eleventh. Lynn responded from her home in Brooklyn and she first spotted the Rescue 2 rig on Fourth Avenue in Park Slope, on its way to the Brooklyn Battery Tunnel. In the tunnel, traf- fic was at a total standstill, and even when Pete Martin stomped on the siren pedal there was nowhere for the cars in front of him to move. The guys waved to Lynn. She had been to their firehouse many times; it was the first firehouse where she had eaten dinner when she got her job as communications director for the FDNY.

In the tunnel Lynn also saw the guys she knew from Squad 1. They were jumping off their own rig and piling onto the Rescue 2 truck, to get in faster. Lynn followed the Rescue 2 truck right out of the tunnel, onto West Street, where it turned left and parked opposite the trade center, near the south footbridge. The guys got off the rig, laughing, joking, a little more tense than usual. She said a few words to the guys, then watched them stride off toward the North Tower.

The chief's aide told Ruvolo he saw them head toward the North Tower. This confirms what Ruvolo and the other men have thought all along, but it makes him feel a little bit different. Before, it was all speculation and he retained, in the back of his mind, the hope, the dream, that he was wrong. Maybe, somehow they hadn't gone into the North Tower at all. He fantasized that they had been injured and unconscious in a hospital in New Jersey, caught on the periphery of the collapse, like Al Fuentes. But now, after talking to Lynn and the chief's aide, Ruvolo realizes that his imaginary scenarios have been just that—fantasies—and nothing more.

He knows where they were going. Of course they went in. There were firemen up there, people trapped in the North Tower. A rescue's job is to follow the firemen and make sure they get out.

Four days after the attack, Phil Ruvolo actually finds a place where the world feels quiet and still. After rain on Friday, the warm September sun has returned and the weary captain goes into a building on Church Street, just across from where the towers used to stand. He climbs up to the roof and surveys the scene below, the mountains of smoky rubble—a shocking sight.

Working down in the smoke, heat, and ash, in the canyons, peaks, and valleys of debris, he finds it hard to fully understand the scope of this disaster. There is too much to focus on down there: he's usually staring straight ahead, trying to make sure he doesn't fall off a narrow I beam, or staring at the pieces of building suspended above his head, hoping they don't fall down on him. In the pile, he can't get the whole disaster site in his field of vision. So today he stands on the roof of the building and looks out at the pile. The building on which he now stands was once in the shadow of the towers, but it now looms over their remains.

Twenty stories in the air, the devastation fills Ruvolo's field of vision. But from here the boundaries of the destruction are visible. He sees two big stumps where the towers used to stand. He sees other stumps where the adjoining buildings were. To the west, the debris goes all the way across West Street. In fact, from up here, he can now see that when he and his men hiked up West Street on the day of the attack, they weren't walking on the street; they were walking on a debris pile that might have been twenty feet tall, with rigs, cars, and big chunks of building underneath their feet. It takes him a few minutes just to absorb the enormity of what he is looking at, the grandeur of these buildings reduced to a giant field of molten hot steel and unrecognizable ashes.

Ever since those first years of studying for the lieutenant's test, Phil Ruvolo has been a diligent student of firefighting and rescue work. He knows all the facts and figures about collapses. If you put a four-story brownstone in front of him, he can give you an idea of how much weight it is bearing, where it might collapse, how the people might ride it out to safety. But when he tries to do the math for this kind of event, the results are impossible to comprehend. Each floor was an acre, and there were 110 floors in each building; that makes 220 acre-sized floors falling as much as a thousand feet to the ground. Knowing what he knows about ordinary collapses and seeing what is left, he realizes that the only people with a chance to survive are the ones who were down in one of the seven basement levels. Yes, occasionally miracles happen; someone manages to defy the odds and ride out a collapse, as Genelle did. But there can't be many like her.

For the thousandth time, he thinks about where his missing guys will be found, about what he would have done if he had been the officer responding to this emergency. He pictures himself climbing stairs, he imagines himself surrounded by terrified people. He calculates that by the time the North Tower came down he'd have made it up at least twenty-five stories, even without an elevator. If they were able to take an elevator part of the way, they'd be even higher. Maybe his guys turned around after the first building came down. But he hasn't heard that an order was given to evacuate. Besides, if they were up that high when the first one came down, they would have made it only part of the way back down before their building collapsed too. Every bone in his body says that those guys would have been climbing fast, looking for victims, unwilling to leave until they knew that every fireman was out of the building. If Ruvolo is right in thinking the guys had made it that high—and he's sure about these men—then it would take a miracle for his men to survive.

But miracles happen very soon after a collapse, not on day five. Most survivors get out quickly. After even one day, the number of survivors drops precipitously. After a week, the chances of being alive in a collapse are an infinitesimal increment above zero. At a much-publicized incident in Turkey, after an earthquake, survivors were once rescued a full week after the collapse. That was one of the jobs that Downey always mentioned when he gave presentations about collapses. But that wasn't two 110-story buildings burning at the temperature of ignited jet fuel, coming down on top of each other. Since Paul and Espo pulled Genelle out, not one person has been found alive—not in the last seventy-two hours.

Now, looking out on this landscape, Ruvolo allows himself to think, for the first time in four days, *They're not coming back*. But only for a moment. It's not something he can bear to say to the other guys or to the wives yet. Maybe he's wrong. Maybe they're alive in there somewhere. But up here, out of the smoke and heat, looking squarely at what lies before him, thinking like a rescue expert and not like a friend, Captain Ruvolo doesn't think there is much left to hope for, except to find the men quickly and bury them properly.

Later that night, back at the firehouse, he makes another round of calls to the wives. Then Ruvolo heads home to Staten Island to see his girls and to let them see him. He needs a change of clothes and a few minutes away from the incredible pressures of working on the pile— a little bit of normal life. Traffic is light, as it has been every day since the eleventh, and he cruises home, first winding through the Flatbush streets he knows so well, then climbing a ramp onto the BQE, which rises above Brooklyn and lets him see Manhattan in his rearview mirror. The skyline now seems empty and sad. On the highway he picks up speed and falls into a meditative state, just as he always does when he's driving home from work. The only thing before him is the road. He just stares at the road.

He thinks about what he saw from the top of that building today, and Phil Ruvolo considers what he and his men are going to face in the days and weeks to come. When he rolls down the window, he can feel the breeze hitting his face. He can smell the asphalt, the diesel, and a little hint of the ocean, not far away now. And he tries to make his peace. He says good-bye to each of his men, right there on the BQE. He shouts his farewells out the window, into the air, to every one of his guys. If the dead can hear him in a church or a cemetery, why can't they hear him on the BQE?

He says good-bye to each man, one at a time. To each he finishes with the same apology: "I have no time to grieve for you now." Maybe he says that because it's true; he has to work at the site, call the wives, run the firehouse. But maybe he says that because he is not ready for these feelings yet; maybe his brain understands what he saw from that building, but his heart cannot take it in. It's still just too much.

BEFORE LONG THE RESCUE 2 guys get their hands on some blueprints of the trade center. The men pore over them, trying to figure out new ways to get deep into the collapse zone. They know the missing guys have to be on the stairs, somewhere. The only real rescues made at the site have occurred on a stairway that was partially intact in the North Tower. That was where 6 Truck was found that first day and where Paul found Genelle on the second day. The problem is that, apart from that stairway and a few others, most of the access routes to the stairs from the top of the pile are closed off by debris. When they realize that hitting the stairs from the top of the pile is impossible, the Rescue 2 guys try to get access to the stairs by going underneath the pile and climbing up.

The next day, September 16, Ruvolo and his men climb down into the tunnel where the PATH train had once run, carrying commuters to New Jersey. They want to get down as deep as they can, then work their way up from there, maybe find a way to access some of the stairwells where more people might be trapped. They enter at a building on Church Street that is still standing, then wind their way down a series of stairwells and passages that take them to the subway level. It's another world down here, absolutely dark, and if their lights fail they will be lost forever. They climb through small holes and passages, seeking the lowest level. Some of these holes lead to nothing; others open onto giant chasms. It is eerie to find an enormous open space after squeezing through such a tiny hole. They hear creaking sounds coming from above and they are in constant fear that at any moment the whole thing might shift and come down on them, leaving them trapped.

Climbing out on a narrow ledge over one of the chasms, the men find a stairway leading to the PATH train tunnel. Bingo! At the bottom of the stairwell they pause. There's a small place they can squeeze through to get down to the train beds. However, the crevice that they push through looks unstable to Ruvolo. He has someone mark it, then leaves Bob there to help guide them out if they get stuck or lost. Bob stands there for a few minutes, his light the only sign of life in the area. Then something falls nearby. Suddenly the crack starts to enlarge. They're going to get caught down here if they don't move out right now.

Bob is already on the radio to Ruvolo. *"Cap, come back to the hole quickly."* Ruvolo can tell from Bob's voice that they are in trouble, and he and the guys hightail it out, crawling back as fast as they can. Bob hears their knees scraping the ground as they approach. Just after they get out, the crack shatters. The opening where they stood five minutes earlier could be flattened by the collapse. On their long climb up to the surface they dub their run back to Bob the Rescue 2 Track Tryout, laughing a little at themselves. The escape brings a sense of relief, but it also reminds every man just how dangerous it is to be hitting these voids right now.

WITH SO MANY of the top brass missing or dead, it takes days to set up a command structure to supervise the thousands of people who stream into the site. At a normal fire or emergency Ruvolo and his men might have, at most, fifteen minutes without one of the big guys taking charge and barking out orders. Not that Rescue 2 always obeys the chiefs, but the chiefs are always there. Right after the collapse, there were a few chiefs who tried to take responsibility, but it was days before there was really any central command structure telling each company exactly what to do.

Phil Ruvolo has to step up and take command of his men as a chief would. Obviously, this isn't like any other operation he has ever seen. In the beginning, when he was a little shell-shocked, he let the guys take crazy, insane risks. They walked on I beams a hundred feet off the ground, where a slight misstep would have killed them. They worked next to molten steel, red with heat, that would have burned right through their bunker gear if they as much as grazed it. They dove into voids that were unstable and looked like they might give way at any moment. This was when Ruvolo still thought they might find some people alive in the rubble. "Take every risk you are willing to take," Ruvolo had said. "Work until you can't work anymore. There are firemen in there and our mission is to get them out."

By the sixth day, Ruvolo has realized that the guys probably aren't alive. Now he changes his approach. He has given the men six days of complete freedom, to hit every void they can, to take every chance worth taking. But now they are starting to hit the same voids, again and again. They aren't finding any new places where people might be. They are just finding new ways to get into the same old voids. The men want to take these risks anyway, to put themselves in danger even when they are hitting the same old void a second or third time.

But Ruvolo now orders his surviving men to stop taking chances. It's just too risky, considering the small chance of finding a survivor.

It's very hard for Ruvolo to convince his men to slow down, pull back. They want to keep going. They are willing to put themselves in any sort of danger. Finally, he has to gather the men together and try to make them understand how precious they are to the city now, how vital their experience is in the case of another attack, and how important it is that they do not die in vain. Like everyone else, he only wants to think about the guys that are missing. But as captain, he has to at least try to think about the future.

"No more feel-good missions," he says. "You're not doing this for yourselves. If you see an opening and no one's been down it, go for it. But don't put yourself in harm's way just to put yourself in harm's way."

Ruvolo still wants to let the men do everything they possibly can to get the guys. Only if they do everything will they be able to look their dead friends' wives in the eye, or look at themselves in the mirror. But at the same time, he has to consider what to do about Brooklyn: His mission isn't just to protect and recover his own men; it's also to protect other firemen and civilians in Brooklyn. What if every Rescue 2 man is working at the pile and a probie dies in a fire in Bed-Stuy? As much as he wants to stay focused on the men he has lost, he can't let down the borough. He needs to make the company functional again in Brooklyn.

When he tells the men they are going back in service in Brooklyn, their tempers flare. Richie sees it as a betrayal of the guys who are missing. Firemen never leave a blaze until every man in their company is accounted for, even the dead. And Rescue 2 always stays until every fireman, from any company, is recovered. In Richie's twenty years in Rescue 2, the company never left a fire while another fireman was still inside, alive or dead. And he isn't about to see his company do anything different now, just because there are more than three hundred dead.

The men are fatigued, living under stress they have never imagined, living through war and grief, but this argument tears them apart inside even further. It is hard to argue against either position. The truth is they are confronting a situation that makes their old codes of honor and ethics impossible to sustain. How do you decide between abandoning your friends, the guys you've risked your life with for years, and protecting other firemen to whom you also feel a responsibility?

Ruvolo listens to Richie and the other men who oppose his plan, but in the end he sticks with his decision. He knows his responsibility to the city, and to people both living and dead, and he summons all his force to drive this point home with the men. There are no chiefs around to fall back on; this is his decision, and his decision alone. He is their captain now. And he feels it now more than ever, because of what has happened, but also because his chief and his house's old ruler, Ray Downey, is gone. He's on his own now. And he realizes, above all, that running a house doesn't mean doing what the men want or even what he wants. It means doing what he thinks is right. So he gives the order: one group of men covers Brooklyn, one group works the pile, and one group stays with the wives of the men who are missing.

They go by the shop and fill an old rig with any tools that are left there. Almost every piece of equipment that works properly was shuttled down to the pile and is still down there, somewhere. So Rescue 2 has to make do with almost no tools and an old broken-down rig. Reluctantly at first, they scrounge up some ancient tools and supplies left behind in the basement, and they take the first, tentative steps toward being normal rescue firemen again.

All the men prefer to be down at the pile. Dealing with the wives is torture. Very soon it becomes apparent that no matter how many errands they run, conversations they have, kids they help out, they can't replace the guys who are gone. There is no way to ease the pain.

And what little they can do begins to feel inadequate as the depth of the families' losses truly begins to set in.

At home, the men try to sleep or relax with their families, but their minds are full of anxiety and grief. Ruvolo wakes up after a few fitful hours of sleep and he can't go back to bed. He can't stop his mind from racing. He asks himself, "What the fuck am I doing at home, with my guys lost at the trade center?" That's when he heads into the firehouse, where he finds others who also can't sleep. There is simply no rest to be found. Often, when their feelings are raw like this, they head down to the pile to do some work.

JOE AND CHUCK DOWNEY go to the site every day to search for their dad. They work it out so that one of them is always at the site. As captain of Squad 18, Joe also has eight of his own men to find, bury, and mourn. He and Chuck talk to everyone who was there when the second tower fell, trying to pinpoint where their dad might be. But it's almost impossible to translate the old geography into this new landscape, and the only thing they find at first is his car, buried over on West Street, with some of his personal belongings inside.

At home, Rosalie Downey and her family pray for Ray. They cling to any hope they can find. One day a report comes on the television that some firemen have been found; immediately, they start to pray that Ray will walk out of there. But he doesn't.

When Ray made chief, seven years ago, Rosalie thought her years of worrying about this kind of tragedy were over. She and everyone else in the family breathed a little easier; to them, the promotion meant that the real risks of the job were behind him. Chiefs still get close to the blaze and often go in to check things out. But they aren't on the front line, at least not in a normal fire or emergency. Before 9/11, a chief hadn't died in the line of duty for over thirty years.

One day Joe Downey is in a van heading north from the site, back to his firehouse. He's with a few men he knows, men who served under his father at Rescue 2. There's also a chaplain in the van who knew Ray from the Oklahoma City operation. He starts talking to Joe about his dad, describing how funny, easygoing, and gentle Ray was in Oklahoma City.

Joe and the other men just nod. Then Joe can't restrain himself any longer. "Father," he says, "are you sure we're talking about the same Ray Downey?" The other men around Joe, who knew his father as a stern commander, crack up. Ray was someone different out at Oklahoma, they find out.

Ray Downey's old friend, Al Fuentes, the man he had shared so much with, lies in a hospital. When rescuers found him, one hour after the second collapse, he was in a state of respiratory failure. Doctors had to induce a coma in order to save his life. Then they gave him an emergency tracheotomy so he could breathe.

At the hospital it was determined that Al had a fractured skull, a collapsed lung, and broken ribs. Now he remains heavily medicated, and the doctors have asked his wife, Ellen, not to tell Al what happened until he asks. He has been in a sleepy, unfocused state. He hasn't asked about anything or asked for anyone.

But about two weeks after the disaster, he wakes up screaming in the middle of the night. He sees jumpers exploding in front of his eyes. His wife rushes to his side and tries to comfort him.

"I gotta talk to Father Judge," Al says. "I think I'm going crazy."

"Father Judge?" she asks. "What do you remember?"

"Well, the South Tower collapsed," he says. "Wait, are you telling me Father Judge is dead?" She nods.

"Ray Downey—is he dead?" She nods again. They hug and cry. Then he asks more names. But soon he just says, "Don't tell me anymore." It's too much.

THE HARD LABOR at the pile is the only thing the Rescue 2 men can feel good about. Their hours there are filled with action—and, inevitably, rebellion. Their friends are dead, their city is in mourning, but that doesn't make these guys any easier to get along with or to boss around. In fact, it makes them much more rebellious, disobedient, and surly; it's as if they want to prove that, even if everything else is gone, they've still got their anger and attitude. The men try to work at the pile whenever they can, no matter what their superiors tell them. Ruvolo's group nicknames themselves the C-Group Volleys because Ruvolo, overwhelmed with all the tasks before him, doesn't fill out their payroll sheets until weeks later. It just slips his mind. So, in addition to everything else, they are working without pay, as volunteers. The C-Group Volleys consist of Ruvolo, Bob, Pete Romeo, Duane Wood, Ray Smith, Bill Esposito, and Vinny Tavella. They listen to Ruvolo, at least most of the time. They can't disobey him—he's their get-out-of-jail-free card.

Like most of the Rescue 2 men, Pete Romeo has problems with authority. He is his own authority, period. He can tolerate a good boss like Ruvolo, or Lieutenant Lund, or Chief Downey; but even before 9/11 he had no patience for some anonymous chief ordering him around. Now he is even less likely to do as he's told. Every order feels like an insult. At one point, he and Bob are walking into the center of the pile and pass a tent marked "Corrections Department." A chief inside the tent yells at them to stop and, as usual, they pay him no mind and keep walking to where they want to search. Before they know what is happening, a soldier apprehends Romeo; then the chief follows, livid and screaming. Bob tries to run interference, explaining, "Sorry Chief, but we were sent on this mission from FEMA. We're not under FDNY authority anymore." But the chief will have none of it; he wants Romeo kicked off the site right away.

Bob gets on the radio to Ruvolo. *"Cap, you better come over here, this chief is doing somersaults."* Eventually, with much finagling on Ruvolo's part, the guys are allowed to go where they want to go. A lot of Ruvolo's time is spent like this, arguing for his men, trying to get them into the most dangerous places, where they think the missing Rescue 2 guys are. The rest of his time, it seems, is spent arguing with the men. But increasingly, they are listening and taking his orders without giving him grief.

The order comes down that every team going below grade has to take a structural engineer with them, to assess the conditions and make sure the area is safe. Ruvolo takes Rescue 2's assigned engineer aside at their first meeting and tells him, "I don't care what we do, you're gonna say it's okay. If you want to come with us, you're welcome. But you're not gonna get on the radio and say it's dangerous."

THEY TRY TO JOKE whenever they have the heart. Some of the best laughs Ruvolo gets come while reading the children's cards being passed around at the site. One reads simply, "Sorry about your friends. I didn't do it." Another card features a crayon drawing of the two towers standing in the Manhattan skyline and the caption "Before." Opening up the card, Ruvolo sees two big piles of flaming rubble. Written at the bottom is: "Better Luck Next Time."

When the men receive gifts of memorial artwork at the firehouse, intended to honor their fallen brethren, they appreciate the support of so many people. But they also find it hard that every aspect of their grief and mourning has become so public.

Every day, on TV, the men see the screen full of names and faces of men they know, and they hear anchor people discussing Ganci, Feehan, and Downey as if they are their old friends. The firemen have built-in bullshit detectors and a low tolerance for the media's portrayal of their heroism. To them it's too Hollywood. Besides, when it is over, their friends will still be dead.

Official counselors are assigned to each firehouse to get the men through their grief. Rescue 2 doesn't get along with its first counselor, or its second one, not the third one or the fourth one, either. Some of the counselors are just shocked at the macabre sense of humor these guys have. Others are scared. One conversation much repeated around the kitchen table goes something like this:

"Are you feeling *angry?*"

"Yeah, come to think of it I am. Listen—how 'bout I beat the crap out of you? I think that would help me."

After one counselor diagnoses many of the men with "misdirected anger," the phrase sticks. It becomes a favorite, sarcastic reply to a vicious outburst in the firehouse. When someone smashes a cake or kicks in a door, the inevitable comment is, "A little misdirected anger, huh?"

The only thing that actually helps the guys is to talk with one another. At first, the men find themselves speaking only lovingly and reverentially of their dead. But then they start to feel stupid—they sound like the guys on TV. They didn't sweet-talk John or Lincoln or anyone else when they were alive, so why start now? They liked the guys who are gone not only because they were heroic, courageous, and every other adjective that's been repeated thousands of times a day on CNN; they also liked the dead guys because they were irreverent wiseasses, relentless ball breakers, mammoth eaters and jokers, and many other things they can't mention to anyone else. So the men start to break balls, even dead men's balls. They make fun of the men who are gone the same way they make fun of the guys still sitting around the kitchen table. And that feels better than anything else. That feels like what the guys would want. Because in the kitchen, honesty is vital. Always has been, always will be.

B Y  T H E  E N D  O F  S E P T E M B E R, the C-Group Volleys make it down to the parking level, where they find some cars still intact. There is a truck from an office-supply store, but no one is in it. The firemen look closely at one of the cars and are able to surmise, from the license plate frame, that it is probably a Port Authority policeman's personal vehicle. Down here, things are just as they had been on September 11 which now seems like ancient history. Everything is untouched, and it's almost possible to imagine finding someone down here, alive, just waiting to be rescued. They know, however, that it is almost three weeks into the operation, and anyone down here would be dead, even if he or she survived the collapse. This is a recovery mission now.

Next to one car is a stairwell that a FEMA team has marked with a warning: "Do not enter, unstable." The Rescue 2 guys just laugh and walk up the stairwell.

They climb toward the lobby level and see some blood on the wall of the stairwell. Next to the bloodstains, a pair of shoes sticks out from between two enormous concrete slabs. The men hope that maybe these are just shoes, with no one in them. The slabs are pancaked together, so they can't even see the tops of the shoes, never mind the legs. They cut open one sole and confirm what they all dreaded: this is a person, crushed by the slabs.

The shoes are steel-toed boots. *The kind a cop would wear,* thinks Bob. Firemen know how to spot a plainclothes cop from across the subway platform. They know how a cop dresses when he's undercover, and what precise uniform each police branch wears. Bob has a powerful feeling that they are looking at the body of a dead policeman.

Now they wonder if the car with the Port Authority police license plate cover belonged to this guy with the steel-toed boots. No dead policemen have been found since the collapses. Thirty-seven Port Authority Police are missing, and twenty-three are missing from the NYPD. So far, the C-Group Volleys have been lucky, if you can call it that, in finding bodies. In three weeks they have recovered many firemen and civilians—just none of their guys. But somehow these two steel-toed boots peeking out of the concrete are the worst. They stop the men cold. There is no way to get the man out right now. They are powerless against these concrete slabs. They just stare at the boots for a little while, then climb up to street level.

Later, when Bob and the men find a Port Authority police lieutenant and a Police Department officer, they tell them that they think they've found a policeman down there. They crawl back in with some police, winding their way down to the garage, then up the precarious stairs to the bloodstains.

As soon as the policemen see the shoes, they know who it is; it's a Port Authority policeman. The guy was only three feet away from safety. When the cops see that Bob is carrying a portable sawzall, his favorite tool, they worry that he is going to try to cut off the feet so the family can bury the guy immediately. Stranger things have happened with people under this much pressure.

Bob sees the way they are looking at this tool. He reassures them. "It's just in case we get caught on something and need to cut our way out." Later, the chief medical examiner comes by to get some DNA for testing. It will take a while to move the slabs and he wants to test immediately and confirm who it is. A minister accompanies him, a guy who had been a fireman in 11 Truck, and they all say a prayer together after they have done their terrible work.

IN LATE SEPTEMBER, Bob and Ruvolo work a day tour in Brooklyn, and they go with the rest of the guys on duty to visit Louis Valentino's dad in Red Hook. They just want to tell him that they haven't forgotten Louis.

"We've got a lot of stuff going on," Ruvolo tells him. "But we remember you too." Louis' dad understands what they are feeling in a way that no outsider can. He has lived with the same feelings for the past five years. Driving back to the firehouse, Ruvolo and Bob hear the dispatcher on the radio telling the captain to call him on the company cell phone for a private message. Then he tells Ruvolo to call Billy Walsh, a lieutenant in SOC, who is working at the pile.

"Phil, we found some of your stuff," Billy says quietly. "We uncovered a mask that looks like a Rescue 2 mask. Maybe you guys want to get in here."

"Okay," Ruvolo says. "We're heading in,"

"They think they might have found somebody," he tells Bob. "They uncovered one of our masks. Let's head to the pile."

Next Ruvolo calls the Brooklyn dispatcher. "Listen, don't call us," he says. "We're not gonna respond."

They turn on the lights and siren and shoot through the streets. They send word back on the intercom to the guys riding in the rear of the rig. This is the moment they've been waiting for, the best thing they can hope for. It's hard to feel good about the chance to find a dead body. All they can feel is relief. Maybe now the wives won't have to wait any longer for a proper funeral.

In Manhattan, Bob drops them off close to the pile and then goes to park the rig. He has to ditch it somewhere far away because they are not supposed to be working at the site; they're supposed to be in Brooklyn.

Ruvolo calls Walsh again and gets the details. "I'm at the north wall of the South Tower, about fifteen feet in," Walsh says. They trudge over the ruins. For weeks they have been searching in the North Tower for the Rescue 2 men, but now someone has been found in what they thought were the remnants of the South Tower. But when Ruvolo looks toward the North Tower from this spot, he can understand; from this vantage point he can see the way that the North Tower, which was the second building to collapse, fell on top of the already collapsed South Tower. They've been looking in the wrong place all along. The men were in the North Tower, but they fell where the South Tower used to stand.

Firemen are digging around what looks like a Rescue 2 mask. When Ruvolo and his guys arrive, the other men stop working and let them climb to the bottom of the hole to work on removing their brother. As soon as Ruvolo sees the knife scabbard attached to the light strap, he knows it has to be Lincoln. Lincoln always had that big Bowie knife across his chest. It's the knife he used at the Atlantic Avenue Fire to cut Timmy Stackpole out. A glimpse at the name inside the turnout coat confirms that it's Lincoln. They dig for a while; the rubble is densely packed around him. When they finally uncover his remains, they lift their friend into a Stokes Basket, cover it in an American flag, and pass him out of the hole.

Steel crane platforms are set up in the area. Ruvolo and three of his men walk Lincoln's body to a crane platform and kneel there next to him, a hand on the Stokes. Somehow, even though they knew this was coming, finding his body makes it real in a way that it hasn't been so far. It was abstract before, but now Lincoln is lying next to them under an American flag. As the men brace the Stokes, a crane hoists them into the air, high up and over the debris. It feels like they are floating out of the rubble, rising into the sky.

The crane places them down in the mezzanine area near Church Street, where all the officials are waiting. Mayor Giuliani is there and he comes over to Ruvolo and says, "I'm sorry, Phil."

Ruvolo just nods. He doesn't know what to say. One of his men. That's all he thinks. This is what it's like to be in command. This is what Gallagher felt, and Downey felt, and now it is his turn, his responsibility to bear. He is the captain.

Bob rides with Lincoln's body to the morgue. They've spent weeks searching for Lincoln, and they don't want to leave him alone now. The other men climb back over the pile to the place where they found Lincoln and start digging further. They figure the guys were probably together when the building came down, so they are hopeful that the rest of the bodies are nearby. After a little more digging they uncover another mask marked "R2," and when they look at the stuff in the coat, they know it has to be Kevin; he always packed his pockets with tools and gadgets. They dig for forty-five minutes to free Kevin's body and are again lifted by crane out over the rubble. Ruvolo, who has known Kevin for what seems like forever, accompanies his friend to the morgue.

Later that day, Ruvolo puts on his Class A dress uniform and travels out to Lawrence, Long Island, with Lieutenant Pete Lund and some chiefs to see Maryann O'Rourke. Ruvolo knows Maryann from Christmas parties at Squad 1, years ago, when she used to bring Jamie and Corrinne to see Santa.

This is the first time Ruvolo has ever had to make one of these calls, but it won't be the last. In the months to come he'll have to make more of these house calls than any single captain had ever made before 9/11. And there are other captains making even more calls than he is.

When Maryann sees the dress uniforms, she knows what it means. For seventeen years this formal wear has meant a funeral. Whenever she saw Kevin putting on his cap and gloves, she knew where he was going. Now, as Diane Valentino had once seen Ray Downey waiting at her door, Maryann sees Ruvolo on her front steps, all dressed up, and she knows why he is here. Ruvolo walks Maryann to the back porch. He sits her down on a patio chair. Then he sits down next to her and says, "We got Kevin."

LINCOLN'S FUNERAL ON SEPTEMBER 27, 2001, is the first Rescue 2 funeral. It catches Ruvolo by surprise. He's never done this before, thank God, and all the pomp and ceremony require that he know where to march, stand, and salute—things he hoped he would never have to learn. In the months to come, Ruvolo will get better at the funerals, and after seven he'll be a pro, but for now it is hard just to know what to do.

Ruvolo marches alone, ahead of the caisson, the fire vehicle that carries the body. Immediately behind the captain, flanking the truck, are his officers, and behind them are the firemen, in neat rows on both sides of the rig. Two men ride on top with the body, one hand on the casket, one held up in salute. Lined up on the church steps are not just the current members of the company, but most of the retired Rescue 2 men as well. They have dug up their old uniforms and come to salute Lincoln. Some of the men look much older now, and their uniforms pinch or hang uncomfortably. But it doesn't matter. The occasion, the history, their devotion to the company, give them all the dignity in the world.

The streets of Sayville, Long Island, are jammed with cars. It seems as if everyone in town is lined up to say their farewell. The church is just a few blocks from Main Street, and the town feels more like a hamlet in the Midwest than a New York suburb. The sidewalks end just a block from Main Street, and people line up in the street or on the grass. The church is tiny, not meant for a huge funeral like this, so only the people who know Lincoln best can squeeze inside. The others listen to the service outside, over a loudspeaker, and stare up at a brilliant blue sky, as cloudless and vivid as it was on the morning of the eleventh. The weather is a little brisker now, not biting like winter, but chilly. Inside, Lincoln's brother, Chuck, a professional singer who works at a resort in the Cayman Islands, gets up and tells stories about their childhood, then sings a song he wrote about Lincoln.

When the mourners leave the church, they are led by the bag-pipers. One man calls the entire contingent of firemen to attention, and hundreds of men salute Lincoln as his casket passes by. Somehow the melancholy sounds of the bagpipes seem appropriate on this cool autumn day. By now, these bagpipes have become so achingly familiar that one note is enough to trigger the tears. Those saluting see only three or four bagpipers, not the full complement. The rest are at other funerals on the same day.

It is small differences like this that make the 9/11 funerals feel different from past line-of-duty funerals. Instead of the usual eight or ten thousand firemen, there are only a few hundred at each funeral.

Instead of the drawn-out get-together, there is a reserved collation at a nearby firehouse. After that, the guys head back to work to look for more of their dead friends.

PHIL RUVOLO HAS TO WRITE a eulogy for Kevin. It hurts to do it. It is the hardest thing he has ever done in his life. He doesn't like to speak in public anytime. But to have to speak now, about this man he loved, is torture. He sits in his office and just stares at the computer for hours, trying to figure out what to say. In the end it all comes together. When he stands up at St. Joachim's in Lawrence, just one day after Lincoln's funeral, he thinks for a minute of Kevin smiling, loving, full of life. And he is able to evoke his friend, through his stories.

For Maryann, returning to St. Joachim's for the funeral means returning to where she first met Kevin when they played together in the church band. At the funeral hundreds of strangers line up, people Maryann has never seen before, but somehow so many of them knew her husband.

Ruvolo gets through Kevin's funeral and through many more—for his men, for his friends, for a cousin from 10 Truck, for men from nearby companies, for men he came up the ranks with. Each of the ceremonies is hard in its own way. There are the guys he knew best, like Kevin and John. Then there are the guys like Billy Lake, whom he was just getting to know. In the last six months, Ruvolo and Billy had been speaking on the phone to each other for forty-five minutes every night. They were becoming good friends. A few weeks after Lincoln's and Kevin's funerals, the men recover the bodies of Billy Lake and Pete Martin. Next to Billy's body they find an elevator door marked "56." Those guys got all the way to the fifty-sixth floor.

By Christmas, they have found everyone but Eddie Rall and John Napolitano. The two families without bodies have it the hardest. The others have a place to be with their loved ones. Darlene and Anne don't even have that. One of their most agonizing dilemmas is deciding when to give up hope of finding their men and hold a memorial service. Eddie's wife, Darlene, is an old friend of Bob's and Bob has discussed this issue with her for hours at a time over the past two months. She decides to have the memorial on November 14, 2001.

Bob will eulogize Eddie. Though Bob is easily the most articulate of the Rescue 2 men and speaks with passion and authority about almost anything, this eulogy frightens him. Over the last two months, writing a eulogy has become even harder than it was before. Bob has been to so many funerals that he feels as though everything that you could possibly say about a fellow fireman has already been said. And he can't just copy what's come before. He has to remember Eddie for everyone and say the last words that will be said about his friend. Bob doesn't show the eulogy to anyone. He quotes from the "Charge of the Light Brigade," which he thinks is fitting given Eddie's courage and combativeness. It's the hardest thing Bob has ever done. The other men say it's the best eulogy they've heard.

The men do everything they can to honor their brothers and help out the girls. But, in many ways, the firemen are robbed of their rituals. The men of the FDNY live by tradition. Their rituals, real and powerful, define life and death for these men. When Louis died, the rituals kept these men whole. And it wasn't just the formal rituals that had been so important; it was also the informal ones, like telling stories around the kitchen table after his death, talking in a bar the night of his funeral, or helping out Diane and the Valentinos. These rituals had reinforced the idea that Louis' death was not forgotten, but honored; that Louis remained forever with them, each time they stepped in the firehouse.

But when it comes time to honor the men who died on 9/11, they simply can't do everything they would have done if there was just one death. When Louis died they put a wooden plaque over his cubbyhole on the apparatus floor. It has remained there, with his turnout coat hanging below, for the five years after his death, and it will remain forever. But if they make memorials out of seven more cubbies, there won't be enough space left for the men who are alive. So many paintings, sculptures, and plaques arrive to honor the dead that Ruvolo has to lay down a rule in the firehouse: any memorial stuff goes up to his office for approval before it goes on the wall. It would be disrespectful to hang something and then take it down, but they don't have enough room to put everything up. As much as he wants to honor the men who have died, he can't let the firehouse become a shrine; guys still have to want to *work* here, not just come to mourn their friends.

The outpouring of public support and sympathy is extreme. There are offers of trips to places all over the world. Lincoln's brother, the singer in the Cayman Islands, invites every man in Rescue 2 down to the resort where he works for a free vacation. For a handful of days the guys sit on the beach, dive with manta rays, and maybe for a second or two, they even forget 9/11.

But Phil Ruvolo doesn't go to the Cayman Islands. God knows they need a break, but he can't feel good yet. Besides, he is needed in Brooklyn.

More than anything else, he and the other men begin to long to just get every last body out immediately, so they can get on with the other work: attending to the girls, taking care of the guys who are left, and having even fifteen minutes just to think about what happened and realize that the men are gone forever.

Ruvolo wants his guys back in fires. They need the feeling of knocking down a blaze or pulling someone out of one. They need to get their asses kicked in a fire; they need to feel exhausted enough to go to sleep.

On November 12, Ruvolo and Pete Romeo are at work, cruising the borough, listening to the police radio as well as their own radio and hoping for a job. A report of a plane crash comes over the police frequency. Ruvolo tells Pete to head toward the crash site in Rockaway, a narrow coastal strip of Queens just south of the Brooklyn mainland. Rockaway is a place filled with firemen's families. Retired lieutenant Larry Gray lives there.

They can see smoke filling the sky as they cross over the bridge to Rockaway, and they can also see fighter planes streaking across the horizon. Ruvolo turns to Romeo. "Pete," he says, "we're all coming home from this one." And they do.

# TEN:

## REBUILDING

E VEN AFTER 9/11, there are still fires that make Ruvolo realize why he stays on the job. In January 2002, Rescue 2 is called to a
fire in a four-story brownstone in Brooklyn, not far from the firehouse.
When the guys arrive, they see flames shooting from the first-floor
windows. The chief on duty tells the Rescue 2 men to check out the
exposure building next door, fast. After parking the rig, Bob slips into
the involved building right behind the engine company. Bob joins the
effort to get the blaze knocked down while Ruvolo and the others are
stuck next door, venting the exposure, opening ceilings, and making
sure the fire doesn't spread to the adjoining buildings.

Eventually, when things seem under control in the exposure
building, Ruvolo enters the fire building, where he is immediately
blinded by smoke. He puts on his mask and moves forward intently,
trying to find the hose. Locating the line, he follows it up toward the
blaze where the engine company is still knocking the flames back. He
wants to get closer. After a few moments, circumstances conspire to
help him. An incessant screeching noise signals that an air tank used
by one of the nearby members of the engine company is nearly empty.
To Ruvolo, the warning sound spells opportunity. He's delighted.
Soon Ruvolo hears signals for a lot of tanks running low. He figures
that he might get a shot at the nozzle as the guys drop out to replace
their tanks. He doesn't have to worry about running out of air himself;

rescue men are equipped with larger air supplies than the men from the engine companies.

The screeching sounds continue, and every time a man leaves Ruvolo gets a little bit closer to the nozzle man. Now he hears Bob nearby, coaching the engine, giving a rah rah speech. "Move in. It's on your right. You got it, bro." He hopes he can sneak up and seize the hose without Bob identifying him.

When the backup finally goes out for air, Ruvolo gets right behind the nozzle man. Then the nozzle man leaves, handing over the hose to Ruvolo. Bob, heavily involved in giving instruction and coaching, isn't paying much attention to who is on the nozzle. He has no idea that it's Ruvolo who's now manning the hose. Bob continues to shout out orders and encouragement, thinking he's still talking to some probie in the engine. "You got it, kid. Bring it in here. Make the turn."

Without saying a word, Ruvolo crawls up in front of Bob, who continues his in-depth instruction. Finally, Ruvolo hears his chauffeur go silent rather abruptly. A rare occurence. Ruvolo knows Bob's finally realized who he's talking to. Gradually, he starts to mumble again, but he's changed his tune. "Uh oh," he says, as if disaster has struck. "Oh boy. I don't like the look of this." Ruvolo knows that both of them are thinking of a similar moment four years back, when they barely knew each other and ended up butting heads in the attic of a burning Queen Anne. So much has happened since then. Now they stand together, two veterans who have been through it all, knocking back this blaze and chuckling under their masks, loving every minute of it.

IN THE MONTHS SINCE SEPTEMBER 11, forty or fifty would-be rescue firemen have made that same scary climb up the Rescue 2 staircase that Bob endured over ten years ago when he arrived to interview with Ray Downey. They have come to meet the captain and vie for spots in the house where so many great firemen have served. Every man who walks in the door knows that Rescue 2 lost everyone working on the day the towers were attacked. To a man, they understand what Ruvolo means when he says, as he now does to

TOM DOWNEY

potential recruits: "There *will* be another attack. We *will* be responding." But not one of those men turns around and walks out the door. They all just nod, meet the captain's gaze, and hope that they will be among those chosen to join the company.

Normally Ruvolo would learn chapter and verse about every man he considers for Rescue 2. But after 9/11 he doesn't have time to scrutinize each applicant. The captain knows that with the added burdens following the disaster—helping out the wives, trying to reequip the company—his veterans will have little time to train new recruits. Instead of taking his new members from strong local truck and engine companies, he looks for guys who have rescue or squad experience and have already undergone extensive training in hazardous materials, scuba, high-angle ropes, and everything else necessary to be a rescue fireman. Ruvolo knows that even if he takes guys with this training, he'll still have to teach them the ways of Rescue 2. But at least they will have the rescue knowledge that an ordinary fireman lacks.

Ruvolo also wants to make sure that, even though he is taking so many new men, the identity of his company holds. He wants to maintain a house that the old Rescue 2 veterans, living and otherwise, would recognize if they walked in the door. The senior men are the keepers of this identity, and Ruvolo knows that there is no one better than Bob to instill a sense of pride in the company and its history in the new men. Bob has become one of Ruvolo's most vital and treasured resources. Since 9/11, their partnership has become even stronger. Ruvolo trusts his friend to keep an eye on the men; he seems to have a sixth sense about what they are feeling. And Bob, being Bob—a talker—tells the men the stories that transform them into true members of the Rescue 2 company, men that Fred Gallagher might recognize as his sort of firefighter. Bob tells them how it was that the mural of the Rescue 2 bulldog came to be painted on the door after the death of Louis Valentino, what it was like for a new guy like John Napolitano to be hazed by Richie Evers, what it was like in the days when Rescue 2 was a house that had never lost a man.

But it's tough to pass on all these stories and keep the traditions alive in a house full of new guys. The company runs the risk of being

280

defined by these new men, not by veterans like Bob. So, in an effort to preserve the spirit of the firehouse, Ruvolo elects to space out the new appointments and brings in a new man only every three or four months. Even so there are conflicts between the new men and the old men.

Bob represents everything old guard. He always sleeps in the kitchen; he smokes like a chimney. He doesn't believe in the firehouse being comfortable or getting remade as a rec room; he likes it just the way it is, the way it always has been. He isn't given to changing things around or cooking heart-healthy meals or pointing the rig toward yuppie fast food. But the new guys are from a different generation. No one smokes. They are healthy and fit. They'll pick a chicken breast over a flank steak. They even want to exercise in the firehouse.

Trouble rears its head when some of the new, younger guys begin a campaign to install a new fitness center upstairs. They've lined up a Stairmaster and some weights for the house, and they've picked out where the stuff should go. The only thing they haven't done is convince Bob, who believes in the privileges of seniority and his own right to be the final arbiter of just about everything. In his mind, his judgment should always stand respected and unchallenged, case closed. Anyone who goes against him or any of the other senior men is bucking up against one of the most ingrained traditions of the job. When the young kids come to Bob and try to persuade him to let them build a workout space upstairs, he just says, "We have weights in the basement. Put the stuff down there."

"But the basement sucks," the guys respond, risking Bob's ire. It's been good enough for twenty years of Rescue 2 men, he thinks—and doesn't budge. He simply refuses to alter his position. Next thing you know they'll want to do yoga around the kitchen table. Then, one day, he comes into work, goes up to the locker room to change, and is greeted by a horrific vision: in the corner is the Stairmaster, tucked away behind a locker. He is not pleased. Storming downstairs, he announces, in his inimitable fashion: "That stuff goes downstairs," but he doesn't get a reaction. "It's going downstairs," he says.

A few minutes later, those gathered in the kitchen hear the loud sound of the Stairmaster crashing down the old wooden staircase, bouncing along the wall on the way down. If they didn't understand Bob before, now they get him. One way or another he will have his way. This is what it means to be a senior man at Rescue 2, and this is what it will always mean.

Still, there are changes. Even though Bob makes a stand on the Stairmaster, some of the new ways of the younger men start to rub off on him and everyone else. Bob actually starts thinking about quitting smoking. And prolonged exposure to the new crop of lean and healthy young men inspires him and almost every one of the veterans to go on protein diets, wolfing down huge heaping portions of meat at every meal but passing on the bread and cake. It is a new day. *Briefly.* For Bob, the diet is one thing. But quitting smoking *forever* is damn near impossible. Bob is used to the comfortable feeling of taking a drag in the kitchen, coffee cup by his side, the *New York Post* spread out in front of him, dropping ashes on the floor because he knows he'll sweep in a little while.

Besides, he does not feel like a new man when he stops smoking. He misses it too much. And what are a few cigarettes after all the stuff he's breathed in during the last twenty years? So he goes back to the butts, then the carbs, and eventually reverts to all of his old ways. Which is, he realizes, just how he wants things to be. He is entitled to his smokes and buttered bagels. He is a senior man.

He feels even more senior surrounded by so many young Rescue 2 recruits. Even a lot of the most experienced new guys are a decade younger than Bob. But, when he sizes these men up, they seem much like the men who are gone.

As THE NEW GUYS become part of Rescue 2, they start to laugh along with the captain and Bob. When they start to joke about somebody in the kitchen, the guy knows he's been accepted. They start to get nicknames. Danny McGuiness is an intense recruit from a truck company. One day he is sitting in the kitchen watching *Black Hawk Down* on the television set, with the volume on full blast. It sounds like the firefight is in Crown Heights, not Somalia. Suddenly a 10-75 comes in over the radio, and the lieutenant bursts into the kitchen and yells to Danny, "Where was that?"

"Mogadishu," Danny yells back.

The next day at dinner when Bob and Ruvolo need the salt or the butter, they address the young fireman: "Hey, Mogadishu, pass the salt." It catches on. Eventually they settle on Mogadishu Dan.

Another new favorite is Sam Melisi, who, besides being a fire-fighter, is also a heavy equipment operator, a professional salvage diver, and a guy who can take apart and rebuild anything from a saw to a fireboat. Ruvolo and Sam have been teaching high-angle rope res-cue together for ten years.

One day Sam catches a project fire. As roof man, Sam's job is to vent the blaze, so as soon as they pile off the rig he streaks up to the roof. The fire is on the top story, the twenty-first story, and Sam takes the elevator most of the way, then walks the last two floors.

The fire floor is raging as he makes the roof. Quickly he takes out the bulkhead, the skylight over the central stairway. But when Sam looks over the side of the building, he sees a woman standing on a bal-cony with smoke pouring out of the window behind her. She's getting ready to jump. Sam is a low-key guy who can barely manage to get excited when he's jammed up in an inferno. The woman down there doesn't faze him. He just yells down, nonchalantly, "Wait a minute." Then he hangs off the ledge of the twenty-one-story building and jumps down to her balcony. He gives her air and calms her down while the guys knock down the fire inside.

"Hey, Spiderman," the chief says when Sam gets down to the street, "what the hell were you doing up there hanging off the roof?" First Mogadishu, now Spiderman. Not a bad start for the new guys, thinks Ruvolo.

All of the new men have seen some fire duty and most of them know the basics of the tools and the equipment. What they have to learn is the way that Rescue 2 operates. Vinny Tavella is a new guy who is used to talking on the radio a lot. When he goes to a job with Ruvolo, he keeps giving bulletins on what he's doing over the handy-talky. *"Rescue roof to Rescue Officer,"* he says, *"I'm opening up the*

*roof.*" Ruvolo hears the transmission and winces. If you're in the rescue, Ruvolo just assumes that you're doing what needs to be done; he
doesn't need to hear it. He only wants his men on the radio if there's
a Mayday or urgent message or something important and unexpected.
Then Vinny transmits another one. *"Rescue roof to rescue officer: I'm
dropping down to the top floor."*

*Rescue officer to rescue roof: shut the fuck up*, thinks Ruvolo, but
he doesn't transmit anything. He just waits for the right moment to
teach the kid a lesson. They walk out of the fire building, change
tanks, and shoot the shit out front, then pile in and drive home. When
they get into the firehouse and hang up their stuff, Ruvolo walks over
to Vinny's coat. He takes the battery off of Vinny's handy-talkie and
sticks it in his coat pocket. Vinny just stares. Then Ruvolo says, "Don't
put the battery back in the radio unless you've got a Mayday. And if
you have time to get the battery on, you don't have one. Trust me."
Vinny gets the point. He isn't heard on the radio for weeks.

BECAUSE THE SPECIAL OPERATIONS COMMAND, Ray
Downey's battalion, suffered such devastating losses on 9/11, there is
a massive shortage of qualified firefighters to fill the ranks of the rescue and squad companies. The men in these companies have to be
specially trained, so the companies can't just take ordinary firemen.
Rescues and squads are particularly important now, because everybody suspects that the next attack will be nuclear, chemical, or biological, and only these companies are equipped to handle these hazards.

What all of this means to the Rescue 2 guys is that, in addition to
all the work at the site and the work training new guys, every man also
has to work the most overtime he has ever worked in his life. In an
ordinary year this would be a bonanza. Guys would be rejoicing over
making a decent salary from firefighting without having to take a second job. But coming right after 9/11, the overtime just adds insult to
injury. The men want a little time to take it easy, but they don't get it.
Eighty- or ninety-hour workweeks are the norm for the whole first
year after the attack.

There is another consequence of this overtime: namely, it gives many exhausted men another reason to look hard at retiring. The FDNY calculates pensions based on the last year of work before retirement, including the overtime worked that year. Many men stand to retire with substantially better pensions if they get out in 2002. No one with twenty years or more under his belt can afford to ignore this opportunity. Besides, for many of the veterans, too much has changed in the FDNY. So much has been lost and the scars are far from healed. They don't want to get over it and move on. Let the rest of the world forget and chatter on about things getting back to normal. To the Rescue 2 men, letting go too soon seems like an insult to the guys who have died. Feelings remain raw.

All the men have a lot to be angry about. A daylong war claimed their friends, their way of life, and much of what had been built in the FDNY for so many years. The experience they lost may never be regained; there just aren't enough fires anymore to give new guys experience comparable to that of the previous generation. Fred Gallagher and his men, Ray Downey and his men, and so many others had built the rescue companies into something no other city had seen. They had reached a state of excellence. Now the life's work of so many has been crippled, and so many good men are gone.

Trying to get back up to speed feels almost like starting from scratch. For men like Bob and Ruvolo, the rescue companies are an accomplishment, an institution just like the towers themselves. Now the rescues are all but flattened too. It's enough to drive you crazy if you think about it too much. And some do. Most, however, soldier on, trying to get through an experience like few men have ever known.

Ruvolo doesn't have much time to think about all these big ideas. He is forced to spend a lot of time up in his office, trying to get a new rig and tools, struggling with department bureaucracy, and doing his best to remember to make sure his men actually get paid. Ruvolo also spends a lot of time talking to the men. In every company, the captain is a counselor to his men. But at Rescue 2, where the guys have given walking papers to four professional counselors, the captain is particularly important: Ruvolo is the only person the guys trust enough to go to for help. He keeps an eye on everyone, asks guys how much they've been drinking, how their wives and kids are doing. He knows that his men are at risk for alcoholism and divorce after everything they've been through, and he does what he can to help them emerge intact.

The transition from 9/11 to normal life is gradual and slow. There's no one moment when Ruvolo walks into the firehouse and thinks things are back to normal. Even two years after, things won't feel normal. Maybe they never will. But after a while the guys start to make fun of themselves, of their own sadness. One day, Mike Quinn cooks some ham steaks, an old-time firehouse favorite. He puts them down on the table and then he says, in mock lament, "I bet Kevin would have *loved* to be eating a ham steak right now. I know John and Eddie *loved* ham steaks too. I bet they would have loved to be tucking into a juicy ham steak right now." But even making fun doesn't make the sadness disappear.

Outsiders often wonder how the firemen comfort each other at this excruciating time. But Bob has no patience for that kind of question. "Comfort each other?" he says in disbelief. "No, that isn't done. There's constant abuse. No comforting each other."

THE WINTER OF 2002 is slow for fires, but even in the down season, they catch a few jobs. After the holidays, Ruvolo and Bob are driving in the rig from the trade center to the firehouse when they hear a job in Brooklyn come in on the radio. Bob is dressed in Carhartt pants and work boots—not bunker gear—but they're passing right by the fire, so they decide to take it in. They make it up to the third floor, crawling past the flames. The place is going good. They're making a search when they run into the safety chief on the top floor. As Bob tries to explain what he's wearing ("Chief, it's a pilot program, new gear"), Ruvolo realizes that listening to Bob bullshit his way through this with all the old attitude is the kind of thing he has missed most in the months after 9/11. Hearing him, anyone in the FDNY would have known he was Rescue 2. It was a signature moment—like the old days.

"Get out of here," the chief roars. Ruvolo and Bob laugh all the way home to the firehouse.

Even with some of the fun of the job returning, Ruvolo and Bob have to think about retirement themselves. They realize that if they have any common sense at all, they will get out now. But then, they joke, who ever gave them credit for having common sense? If they had common sense how did they end up as firemen? Bob doesn't take the possibility of retirement seriously at all. His wife, Linda, may dream about a retirement home in Delaware, but he doesn't think about it very long. This kitchen is where he belongs. Though he makes a half-hearted attempt to study for promotion, he wants to stay in the company. Senior man and chauffeur at Rescue 2 is as fine a place as any, he believes, to end up. He asks Linda for three more years, but he figures he'll renegotiate with her when the time comes. He wants to serve thirty.

Ruvolo has had a lot on his hands. In the month before 9/11 he almost lost his father to an illness. Now every time his dad sees Ruvolo, the older man asks, "When are you gonna retire?" It's easy for Ruvolo to answer his dad, who himself put in thirty-three years in a Brooklyn engine. Ruvolo knows that for most of his dad's last thirteen years on the job, people were bugging him to retire.

"It's not time yet," he tells his dad, explaining that he has unfinished business at his company—to return the house to what it was. Yet Ruvolo also is forced to consider what his family would do if something happened to him. He pays over two thousand dollars a year for the best life insurance policy he can buy, which will give his family a million bucks if he dies. Even though they would like to see him out of danger, they understand that he is a fireman. He doesn't have any idea of what else he would want to become. Gradually, he convinces his wife and daughters that he cannot retire. But he is not surprised when other members of the company come to a different decision.

Pete Lund, Vulcan the Fire God, had always hoped to put in thirty years before hanging it up. He and Billy Lake, his best buddy and driver at Rescue 2, had made a pledge that they would do eight more years together. The night before the eleventh, Billy celebrated his twentieth anniversary on the Fire Department by cooking up a nice meal for everyone, his treat. On the morning of the eleventh, he was one of the original men dispatched to the towers. When he didn't come back, Lund vowed to get through the eight years for Billy. But even Lund can't summon the strength. The job has never been the same for him since the loss of Louis Valentino. Now, with seven more good friends gone from Rescue 2, Lieutenant Pete Lund decides to retire. Another Rescue 2 legend is hanging up his helmet forever.

Promotions also take men away. On the first lieutenant's exam after 9/11, nine Rescue 2 guys pass the test. Their names are put on the lieutenant's list and they are called up for promotion according to their score on the exam. The first wave of promotions takes away five men, including Mark Gregory, Lincoln Quappe's beer-drinking and commuting buddy. By 2004, after all the promotions come through, only seven of the pre–September 11 Rescue 2 men are still in the company. The chief's test is approaching and, with some time at the books, Ruvolo is a shoo-in for promotion. But he doesn't crack the books. He's not looking to move up any further. He's the captain of Rescue 2. What more could he want?

The promotion to chief does bring more authority and responsibility. But Ruvolo thinks back to the fourteen years Ray Downey spent at Rescue 2 before he made chief. Ruvolo wants to put in serious time here and do everything he can with the company.

ONE NIGHT AT A STORE FIRE in Bushwick Rescue 2's men are working to gain access to an involved building. An enormous metal shutter is down, so they attempt to rip it open with a saw and tools. Fire is shooting out everywhere when the shutter contraption finally falls, crashing to the ground. But Pete Romeo's legs are crushed underneath it. His brothers bundle him into a stretcher and give him a cigarette to smoke while he waits for medical help to arrive. They put him on the cell phone with his wife so she knows he's okay.

Ruvolo gets the call at home at one in the morning. Throwing on some clothes, he hops in the car and heads for Bellevue, to an emergency room he knows far too well. Even at one in the morning, the place is hopping. The guard waves him through.

Ruvolo finds Pete in a corner somewhere, waiting for his X-rays to come back. "Take me out for a cigarette" is the first thing out of Pete's mouth. Ruvolo pulls up the brake on the gurney and wheels him through the emergency room to a back entrance where the ambulances pull up. Pete lights a cigarette. His legs look bad.

When Ruvolo wheels him back in, he tells the captain that he wants to go home. The doctors aren't happy about that. He needs to stay, they tell Ruvolo. But Ruvolo wheels him out again, stuffs him in his car, and drives him back to Staten Island.

Pete was one of Ruvolo's first picks for Rescue 2. He's become one of the captain's best friends in the place. It's hard to see him go. But when he finds out how badly he is injured, Pete has no choice— he has to retire or work a light-duty post. And Pete isn't a light-duty kind of guy. He gets out on three-quarter salary but refuses further painful treatments. He's not one for operations. He'll live with the pain instead.

The accident is another reminder that firefighting is a young man's game. It's tough to stay ahead of all the injuries that can end a career. After 9/11, Ruvolo is keenly aware of the whole job, not just Rescue 2, getting much younger. By the beginning of 2003 more than half of the firefighters in the FDNY will have less than five years' experience. That means that half the department did not even start work until 1998, when the amount of major fire duty had dwindled, even for the busiest companies. For the Fire Department and Phil Ruvolo, it seems like a time when the changes will not stop coming. And this new era brings other, enormous complications.

AFTER 9/11, it is clear that Ruvolo and his men are no longer just firemen or even just rescue firemen. In fact, by the time Ruvolo retires, knocking down blazes may be all but ancient history for a house like Rescue 2. Now, when he thinks about the future of the job, Ruvolo doesn't think about fires, collapses, or people stuck under subway cars. He thinks about a dirty bomb going off in midtown, a chemical attack on the subway, a nuclear event in the middle of a busy workday. Ruvolo knows that he must move forward with Ray Downey's transformation of the rescue companies; he must begin to think about the challenges that the next generation of terrorists will bring to cities like New York.

Ruvolo has to rebuild a company that is prepared not only for old-school fires and rescues, but also for stuff that no one has even started to think about. And that's a tall order. He knows that the FDNY no longer has as many experienced men. He has to familiarize himself with all kinds of new threats and he begins the process. He starts to memorize radiation numbers and familiarizes himself with every new chemical and biological agent they might face. Forget the chief's exam. This is his duty now.

Some of the guys aren't so sure they want to be fighting these new kinds of battles. They signed up to be firemen, not chemists or soldiers. Ruvolo calls it the "this wasn't in the brochure" mentality, as in, all this chemical and biological weapons stuff wasn't in the brochure when I signed up for the job. "It's in the brochure now," he tells the guys. "And if you don't want to do this kind of work, this isn't the place for you."

If New York is hit with a nuclear or biochemical attack, Ruvolo knows that a good percentage of men in the FDNY line units will go down in the initial response. Rescue 2 will have to deal with the aftermath, with the men who survive—or don't. He and his five men will be powerless to save everyone, but if they can save a few firemen they will have performed their special mission.

Conversations in the kitchen shift back to topics that the men wouldn't have touched even months earlier. One day Ruvolo comes in, gathers everybody around, and starts a discussion. "What do you do if you have a fireman down and when you hit the hallway to get him your Geiger counter starts chirping like crazy?" he asks. "He's right on top of the radiation source. What do we do?"

"I would go for him," a new recruit tells Ruvolo.

"What if you're just going to end up dead in the hallway too?"

The men bat the subject back and forth. It's one of those questions they will only be able to answer when they are looking down the hallway at that fireman. Before 9/11, before they had seen what could happen if they went in to save people the way they always did, the answer would have been clear: go for him no matter what. But terrorism changes things. They're dealing with more than accidents. The threats to the men are more abstract—and more sinister—and much more lethal.

In January 2003, Ruvolo and his men catch a good job during a blizzard. With ten inches of snow on the ground, it's slow going for the rig. The only good thing is that the streets are deserted. Finally they pull up to the scene at a store on Nostrand Avenue. Ruvolo and Ray Smith get on their hands and knees and crawl in behind another officer, who is standing. But the guy in front doesn't sense that there is a hole. All of a sudden he falls through, plunging into the basement. Ray dives in after him and eventually pulls him out. Then they move in to fight the flames.

Everything that can go wrong does. The hose freezes up just as the engine company is ready to knock down the fire, and Rescue 2 has to drag out two other firemen who are overcome by the heat. The place is filled with acetylene tanks that periodically explode and make them dive for cover. But even with all this stuff going wrong, after everything he's seen in the last two years, Ruvolo just thinks to himself, "This is nothing. This is easy. Hell, this is fun." He may have little patience for the "not in the brochure mentality," but it sure feels good to fight an old-fashioned fire. He didn't sign on to fight terrorists or to be the first line of casualties in anybody's war. He signed on twenty-five years ago to do this.

THE 2003 BLACKOUT, arriving on a hot summer's day, takes New York by surprise. The blackout shuts down the whole of the eastern seaboard and beyond. The last big one was in 1977, back when Gallagher's boys were running all across Brooklyn to knock down the flames. This blackout is much different. There is little crime, almost no looting, and few people get hurt in the city. But the fires burn like the old days. Not like the blackout of '77, but like a good fire night in the late seventies or early eighties, a night with ten jobs.

After the guys on the night tour have a night to remember—twenty-one jobs in fifteen hours—Ruvolo comes in for the day tour and he catches ten jobs that day. The fires pop up everywhere. The company shoots over to Williamsburg for an all-hands and then three more all-hands come up while they're at work. Cruising back to Bed-Stuy, they hear a 10-75 go out. Ruvolo radios in to the dispatcher, *"Rescue 2 to Brooklyn. We're just around the corner from that 10-75—we'll take it in."*

It is a day of fighting fire, a day of redemption. A day where no one gets killed, but everyone has fun. The new guys get some growth. The veterans back them up. The episode reminds a group of men that badly need to be reminded just why they do this job. This is why they signed on in the first place. And it's what no civilian can ever understand. The fire is a pure high, a rush that any self-respecting suburban dad should be ashamed of craving. But the high, the rush, the fun of it are the firemen's secret. It's what they love almost as much as their wives, their kids, their brothers on the job.

That's what Ruvolo continues to think all the way back to the firehouse, after his big day. Then he gets back to quarters and walks past the small memorial wall they've put up outside the kitchen, as he does every time he walks into the firehouse. A door from the old Rescue 2 rig, the one that got destroyed on 9/11, is up on the wall. Kevin O'Rourke, Lincoln Quappe, John Napolitano, Eddie Rall, Billy Lake, Danny Libretti, and Pete Martin; each one of these men is pictured. Ray Downey's name is on top of the list of the Rescue 2 men killed in action. When Ruvolo sees the names of all these dead friends on the wall of his firehouse, he can't think it's just about the rush or thrill of the job. Ruvolo's mission really matters. He owes it to Ray Downey and the rest of those guys up on the wall, and to everybody out there who might need their help, to build Rescue 2 into the men who will protect New York's firefighters in the future. And that's why he comes to work every day at Rescue 2.

# EPILOGUE

AT RESCUE 2 THERE ARE MANY new faces around the kitchen table. To me, everybody looks younger, even the old guys, because they've lost weight and started working out more. But every time I think things have changed, I am reminded of how much they remain the same. The new guys like Mike Travers bust my balls just like the men who are gone. Mike's candor, exemplified by comments like, "What the fuck are you doing here again?" remind me of the same kind of rough but sincere reception I got from Eddie Rall and Lincoln Quappe three years ago when I first set foot in the kitchen.

Bob cooks up linguine with white clam sauce. While everybody winds strands of pasta onto their forks, he launches into a story from his old days in the Tin House. That night, he tells of charging into a fire right behind a senior guy, John Kilgus, who was knocking the flames down with the can in the front hallway. Bob lost a new cap on one of his teeth while putting on his mask. He had just paid $350 for the cap, and he wasn't ready to resign himself to its loss. So he climbed over John's legs, frantically searching for the cap on the hallway floor of the fire building, oblivious to the flames raging around them.

When John tried to back up, he couldn't—Bob was on top of his legs. Now John was starting to get cooked by the fire. He hollered and hollered, but Bob didn't notice. He was too focused on finding his cap to hear anything. Finally the officer piled into the hallway, saw what was happening, grabbed Bob from behind, and tossed him out the door. Unfortunately Bob landed on the spiky point of a Halligan tool. He was impaled. When he stood up, it hung from his rear end like the tail on a bunny rabbit. But this bunny rabbit was screaming in pain. Later, at the hospital, he had to get twelve stitches, but he had found his cap.

Bob's story keeps everyone entertained for the meal. It's the same kind of tale that guys always tell about themselves—something self-effacing that lets the other guys make fun of them a little bit. At Rescue 2 nobody ever pounds their chest and tells heroic stories about themselves; you're only allowed to tell those kinds of stories about other people.

After the meal one of the newest guys, Charlie, sweeps and mops the kitchen floor, and everybody evacuates to let it dry. The guys stand around in the shop area, taking turns painting panels that they hoist onto a workbench. Vinny Tavella comes in after taking out the trash to the Dumpster across the street and reports that two guys out there are cursing each other out and getting ready for some gunplay. He heard them pop a few caps before he ran back into the firehouse.

Captain Ruvolo teases Charlie, the new guy, in his brutally direct way, but the rookie doesn't say much. Maybe that's just the way he is—or maybe he doesn't want to give the other guys any rope to hang him with. When Ruvolo calls Charlie a motormouth, the other men chuckle.

What I realize is that even though the new guys are younger and less experienced, even though they haven't been through the war years or anything like them, even though they grew up in the suburbs and not the old working-class neighborhoods of the city, they are, in the important ways, the same as the men who came before them. They've managed to resist the lure of money. They've chosen to do something they love. They want to serve others for the same reasons that Ray Downey, or Kevin O'Rourke, or John Napolitano did. They are not looking to get their names in the paper. They want to save lives, take risks, win the respect of their brothers.

The rescues lost a lot on 9/11, but they didn't lose everything. Ray Downey and many men of his generation are gone, but there are still veterans to teach the new kids. Downey's son, Joe, is now an SOC chief. Ruvolo is the most senior captain in SOC, and he's committed to remaining at Rescue 2 until he retires. The men are better trained now to respond to a terrorist attack than they ever have been. They go to courses at the Rock every few months on whatever the newest terrorist threat is.

Now, in the kitchen, the guys are as likely to discuss hypothetical terrorist attacks as they are a good fourth alarm. That doesn't mean they won't be scared when they get the call for the next big one. And it doesn't mean they won't think about their wives and children as they ride toward the alarm. But they've considered what might happen and learned what they can about how to protect themselves—and save others.

Tonight the guys go outside and train for a while on a new collapse rig. It's after midnight when they finally pack up the collapse rig and drive it over to its home at a nearby firehouse. Even after they return, Ruvolo sits down in the kitchen with them, drilling the new kids on tides, radiation numbers, and collapse statistics. It is a never-ending job, and still, at this point, it's a fast game of catch-up.

The next morning, after a night with only one crappy false alarm to interrupt a good night's sleep, there's a fire down in Crown Heights. It comes in the last few minutes of the day tour, and everyone hops on with the hope that this job will make the past fifteen hours worthwhile. Ruvolo and Vinny Tavella walk into a giant warehouse filled with smoke and find a truckie pounding on a thick steel door, nearly killing himself trying to pry open the lock. They sneak to the right, pop an easier door, and snake down a hallway. They are now behind the door the truckie has been trying to open. Just after he comes down with another tremendous blow, still not enough to break the lock, he looks through the window and sees their grinning faces. He can't believe they beat him into the fire.

Upstairs the men take a pounding, trying to navigate through piles of goods and supplies in the enormous warehouse. When they finally come outside after putting it out, their faces are covered in black ash. But each man is smiling. Back in the kitchen, they laugh about popping the door before that other guy, then do a critique. "I should have had us deploy a search rope upstairs, to make sure we didn't get lost in that space," says Ruvolo. The young guys nod, taking it in.

Later, Ruvolo stays upstairs doing paperwork, even though his tour is over. The day tour takes in a fourth alarm in Williamsburg. When they return a guy comes clambering up the stairs to tell him that Bob's in Bellevue Hospital with a bad shoulder and back. The pain was so bad that Bob could hardly breathe. Ruvolo drives over immediately to see his friend. A dislocated shoulder, a bad back, these he can deal with. It's just part of the captain's job, the kind of concern that will fill his life as he labors to build Rescue 2 into the best firehouse in the world. Like his predecessors, he is a determined man.

# ACKNOWLEDGMENTS

A third of the author royalties from this book will be given to related charities—the Chief Ray Downey Scholarship Fund and the Rescue 2 Memorial Fund.

This book has only been possible with the help and cooperation of many Rescue 2 firemen who gave me incredible access to their lives, their memories, and their ideas. The men of Rescue 2 have been unremittingly kind and generous to me. That doesn't mean they haven't given me hell every step of the way, but I understand that's the price I pay for setting foot in the firehouse.

My uncle, Ray Downey, made possible the documentary shoot out of which this whole project began. He helped me get my foot in the firehouse door and he gave me advice and continuing assistance. I hope this book honors him. May he rest in peace.

Phil Ruvolo and Bob Galione have put up with countless late-night interrogation sessions at the firehouse, when what I had billed as one last question somehow managed to stretch until two or three o'clock in the morning. These two men have opened their lives to me and made me feel at home in their firehouse. Since they are the most prominent living people in this book, I'm sure they're going to take a beating in the kitchen for letting their names appear in print. I wish I could take that beating for them, but what I can say in their defense is that any errors or omissions are my fault, not theirs.

The men of the seventies, especially Fred Gallagher, Tom Dillon, Jack Pritchard, Pete Bondy, Larry Gray, and Marty McTigue, helped bring to life a world I had only known as a child. These men lived through an incredible time for New York City firefighting; they are our city's equivalent of a "greatest generation."

The men of the eighties helped fill me in on those exciting years at Rescue 2. Mike Pena opened up his memory and his huge collection of Rescue 2 material to me. Especially helpful were John Barbagallo (who is now up there with my Uncle Ray, on the Rescue 2 rig in heaven), Richie Evers, Bruce Howard, Pete Lund, Bob LaRocco, Terry Coyle, Jim Ellison, Al Fuentes, Al Washington, and Sonny Cataldo. Warren Fuchs, a Brooklyn dispatcher from this generation of men, shared his encyclopedic knowledge of New York firefighting with me; he was also kind enough to open his incredible collection of fire photographs to me. These men fought nightly battles to defend our city, and the fact that so many of them are now among the walking wounded is a testament to their sacrifices.

The last generation of men, from the nineties to the present, are the ones I know best. They helped me to understand the contours of daily life at the firehouse and put up with an extra mouth to feed at the kitchen table for a very long time. The list of those men who are now gone is long: Pete Martin, Billy Lake, Danny Libretti, John Napolitano, Lincoln Quappe, Kevin O'Rourke, Eddie Rall, and Timmy Higgins from Rescue 2. Terry Hatton, Dennis Mojica, Joe Angelini, Bill Henry, Pat O'Keefe, Dave Weiss, Gerry Nevins, Mike Montesi, Gary Geidel, Ken Marino, and Brian Sweeney from Rescue 1. Patty Brown, who was captain of Ladder 3, originally inspired me to work on this book by sharing the incredible story of his rope rescue in midtown. May those men and all 343 rest peacefully in heaven, though I'm sure they're already giving one another hell.

The men who survived 9/11 have labored long and hard after the disaster to honor their brothers. They were as kind to me after everything they went through as they were before. From Rescue 2: Sal Civitillo, Mark Gregory, Joe Jardin, Pete Romeo, Jimmy Sandas, Ray Smith, Paul Somin, Duane Wood, Jimmy Kiesling, Billy Eisengrein,

Jim Jaget, Mike Quinn, Tony Tedeschi, Tommy Donnelly, Cliff Pase, Stan Brzezinski, and Dave Arciere. From Rescue 1: Jack Flatley, Paul Hashagen, Al Benjamin, Joel Kanasky, Kevin Kroth, Thor Johannessen, Dave Marmann, Steve Turilli, John Weisheit, Frank Fee, and Glen Bullock.

The families of the firemen who died, especially Maryann O'Rourke, Ann Napolitano, John Napolitano Sr., Jane Quappe, and Diane Valentino, have been incredibly open in remembering their loved ones with me and in talking about events that are painful to discuss.

Frank Campesi and Brian Baiker shared their stories with me. Dr. Artie Weisenseel brought back the lost world of Irish Woodside and the young life of my Uncle Ray.

Tony Horn was the first one to believe in the documentary film that was the genesis of this book. The FDNY press office and especially Francis Gribbon, David Billig, and Chief Brian Dixon have been very cooperative on this project.

George Hodgman, my editor at Henry Holt, went far above the call of duty. His editorial advice and guidance and his unceasing labor have made it possible to write this book. Jennifer Barth and John Sterling have been hugely supportive. Victoria Haire, Kenn Russell, and Chris O'Connell have been very patient in the last, frantic months of writing and their hard work deserves my thanks.

My agent, Heather Schroder, recognized the importance of this story long before the firemen became headline news. She had me working on this book in July 2001; her wise counsel and astute criticism have been much appreciated.

My family and friends have seen me through the long and difficult process of writing this book. My parents, Tom and Eileen Downey, have supported me every step of the way in this and every other endeavor in my lifetime. I am very thankful for their love and counsel.

For as long as I can remember, my cousins Mike and Rich have been like brothers to me. They have supported me and helped me with this and every other project.

My Aunt Rosalie, and my cousins Joe, Chuck, Ray, Kathy, and

Marie, have all been very helpful and supportive in this project. My conversations with Chuck and Joe about their jobs were what originally inspired me to undertake this project. Aunt Rosalie has shared with me the wonderful memories she has of Uncle Ray.

Joy lent emotional and intellectual support throughout and gave me just the right measure of boundless encouragement and earnest criticism.

Catherine Saint Louis made this all happen by publishing Patty Brown's story back in June 2001 and has been a great friend and wise reader. Bill Harrington, Marcus Ryu, and José Marquez have been earnest and helpful readers of this manuscript and wonderful friends. Tom McGeveran has given very helpful feedback. Maura Fritz has offered valuable comments and wise guidance.

Writing this book has been a wonderful, if also sad and painful, experience. But more important than the writing has been the opportunity I have been given to see the way that these men live their lives. I have experienced something real and powerful that has shaped me. And I am forever indebted to all the people who have helped me to see this world.

# ABOUT THE AUTHOR

Tom Downey is a writer and a filmmaker who grew up in a family of firefighters. His writing has appeared in *The New York Times Magazine*. He has worked all over the world producing and directing videos for the Soros Foundation and has taught at a film school in Singapore. He is the nephew of the late Chief Ray Downey, former head of Special Operations Command for the FDNY, who arranged for Tom to live and work with Rescue 2 in order to make a documentary film (which aired on the Learning Channel). Downey spent more than a year at the house before 9/11, and continued to research this book at Rescue 2 after the disaster.